HIGH-PERFORMANCE
MEDICAL IMAGE PROCESSING

Biomedical Engineering: Techniques and Applications Book Series

HIGH-PERFORMANCE MEDICAL IMAGE PROCESSING

Edited by
Sanjay Saxena, PhD
Sudip Paul, PhD

APPLE ACADEMIC PRESS

First edition published 2022

Apple Academic Press Inc.
1265 Goldenrod Circle, NE,
Palm Bay, FL 32905 USA

4164 Lakeshore Road, Burlington,
ON, L7L 1A4 Canada

CRC Press
6000 Broken Sound Parkway NW,
Suite 300, Boca Raton, FL 33487-2742 USA

2 Park Square, Milton Park,
Abingdon, Oxon, OX14 4RN UK

© 2022 Apple Academic Press, Inc.

Apple Academic Press exclusively co-publishes with CRC Press, an imprint of Taylor & Francis Group, LLC

Library and Archives Canada Cataloguing in Publication

Title: High-performance medical image processing / edited by Sanjay Saxena, PhD, Sudip Paul, PhD.

Names: Saxena, Sanjay, 1986- editor. | Paul, Sudip, 1984- editor.

Series: Biomedical engineering (Apple Academic Press)

Description: First edition. | Series statement: Biomedical engineering: techniques and applications book series | Includes bibliographical references and index.

Identifiers: Canadiana (print) 20210390018 | Canadiana (ebook) 20210390050 | ISBN 9781774637227 (hardcover) | ISBN 9781774637333 (softcover) | ISBN 9781003190011 (ebook)

Subjects: LCSH: Diagnostic imaging.

Classification: LCC RC78.7.D53 H54 2022 | DDC 616.07/54—dc23

Library of Congress Cataloging-in-Publication Data

....

CIP data on file with US Library of Congress

....

ISBN: 978-1-77463-722-7 (hbk)
ISBN: 978-1-77463-733-3 (pbk)
ISBN: 978-1-00319-001-1 (ebk)

BIOMEDICAL ENGINEERING: TECHNIQUES AND APPLICATIONS BOOK SERIES

This new book series covers important research issues and concepts of the biomedical engineering progress in alignment with the latest technologies and applications. The books in the series include chapters on the recent research developments in the field of biomedical engineering. The series explores various real-time/offline medical applications that directly or indirectly rely on medical and information technology. Books in the series include case studies in the fields of medical science, i.e., biomedical engineering, medical information security, interdisciplinary tools along with modern tools, and technologies used.

Coverage & Approach

- In-depth information about biomedical engineering along with applications.
- Technical approaches in solving real-time health problems
- Practical solutions through case studies in biomedical data
- Health and medical data collection, monitoring, and security

The editors welcome book chapters and book proposals on all topics in the biomedical engineering and associated domains, including Big Data, IoT, ML, and emerging trends and research opportunities.

BOOK SERIES EDITORS:

Raghvendra Kumar, PhD
Associate Professor, Computer Science & Engineering Department,
GIET University, India
Email: raghvendraagrawal7@gmail.com

Vijender Kumar Solanki, PhD
Associate Professor, Department of CSE,
CMR Institute of Technology (Autonomous), Hyderabad, India
Email: spesinfo@yahoo.com

Noor Zaman, PhD
School of Computing and Information Technology,
Taylor's University, Selangor, Malaysia
Email: noorzaman650@hotmail.com

Brojo Kishore Mishra, PhD
Professor, Department of CSE, School of Engineering,
GIET University, Gunupur, Osidha, India
Email: bkmishra@giet.edu

FORTHCOMING BOOKS IN THE SERIES

The Congruence of IoT in Biomedical Engineering: An Emerging Field of Research in the Arena of Modern Technology
Editors: Sushree Bibhuprada B. Priyadarshini, Rohit Sharma, Devendra Kumar Sharma, and Korhan Cengiz

Handbook of Artificial Intelligence in Biomedical Engineering
Editors: Saravanan Krishnan, Ramesh Kesavan, and B. Surendiran

Handbook of Deep Learning in Biomedical Engineering and Health Informatics
Editors: E. Golden Julie, S. M. Jai Sakthi, and Harold Y. Robinson

Biomedical Devices for Different Health Applications
Editors: Garima Srivastava and Manju Khari

Handbook of Research on Emerging Paradigms for Biomedical and Rehabilitation Engineering
Editors: Manuel Cardona and Cecilia García Cena

High-Performance Medical Image Processing
Editors: Sanjay Saxena and Sudip Paul

IoT and Cloud Computing-Based Healthcare Information Systems
Editors: Anand Sharma, Hiren Kumar Deva Sarma, and S. R. Biradar

The Role of Internet of Things (IoT) in Biomedical Engineering: Present Scenario and Challenges
Editors: Sushree Bibhuprada B. Priyadarshini, Devendra Kumar Sharma, Rohit Sharma, and Korhan Cengiz

ABOUT THE EDITORS

Sanjay Saxena, PhD

Assistant Professor, Department of Computer Science and Engineering, International Institute of Information Technology, Bhubaneswar, India

Sanjay Saxena, PhD, is an Assistant Professor in the Department of Computer Science and Engineering, International Institute of Information Technology, Bhubaneswar, India. He earned his PhD from the Indian Institute of Technology (BHU), Varanasi, India, in High-Performance Medical Image Processing. Also, he completed postdoctoral research at the Perelman School of Medicine, University of Pennsylvania, USA, and worked on brain tumor (glioblastoma) segmentation and analysis. He has published several research papers in peer-reviewed international journals and conferences. He is a professional member of IEEE, ACM, New York Academy of Science, IAENG.

Sudip Paul, PhD

Assistant Professor, Department of Biomedical Engineering, North-Eastern Hill University, Shillong, India

Sudip Paul, PhD, is currently working as an Assistant Professor and Teacher In-Charge in the Department of Biomedical Engineering, School of Technology, North-Eastern Hill University (NEHU), Shillong, India. He completed his postdoctoral research at the School of Computer Science and Software Engineering, The University of Western Australia, Perth. He received one of the most prestigious fellowship award (Biotechnology Overseas Associateship for the Scientists Working in the North Eastern States of India: 2017–2018 supported by Department of Biotechnology, Government of India). He received his PhD degree from the Indian Institute of Technology (Banaras Hindu University), Varanasi, with a specialization in electrophysiology and brain signal analysis. He received first prize for in the Sushruta Innovation Award 2011, sponsored by the Department of Science and Technology, Govt. of India, and he also organized many workshops and conferences, out of which most significant are the IEEE International Conference on Computational Performance Evaluation 2020; 29th Annual Meeting of the Society for Neurochemistry, India; and IRBO/APRC Associate School 2017.

Dr. Sudip has published more than 90 international journal and conference papers and also filed four patents. He completed 10 book projects, and two are ongoing as editor and two are authored. Dr. Sudip is a member of different societies and professional bodies, including APSN, ISN, IBRO, SNCI, SfN, IEEE, IAS. He has received many awards, especially the World Federation of Neurology (WFN) traveling fellowship, Young Investigator Award, IBRO Travel Awardee, and ISN Travel Awardee. Dr. Sudip also contributed his knowledge in different international journals as an editorial board member. He has presented his research accomplishments in the USA, Greece, France, South Africa, and Australia.

CONTENTS

CONTRIBUTORS

J. V. Bibal Benifa
Indian Institute of Information Technology, Kottayam, Kerala, India, E-mail: benifa.john@gmail.com

Rupsa Bhattacharjee
Center for Biomedical Engineering, Indian Institute of Technology Delhi, New Delhi, India

Channa Basava Chola
Indian Institute of Information Technology, Kottayam, Kerala, India

Suchismita Das
International Institute of Information Technology, Bhubaneswar, Odisha–751003, India

Yaman Dua
Department of Computer Science and Engineering, IIT (BHU), Varanasi, Uttar Pradesh, India, E-mail: yamandua.rs.cse18@iitbhu.ac.in

Biswajit Jena
International Institute of Information Technology, Bhubaneswar, Odisha, India, E-mail: biswajit310@gmail.com

Lavanya B. Koppal
Department of Computer Science, Dayananda Sagar University, Bangalore, Karnataka, India

G. Chenchu Krishnaiah
Department of ECE, GKCE, Sullurupeta, Andhra Pradesh, India

Pradeep Kumar
Department of Electronics and Communication Engineering, National Institute of Technology Patna, Bihar, India

Vinod Kumar
Department of Computer Science and Engineering, IIT (BHU), Varanasi, Uttar Pradesh, India, E-mail: vinod.rs.cse18@iitbhu.ac.in

Martin A. Lindquist
Department of Biostatistics, Bloomberg School of Public Health, Johns Hopkins University, Baltimore, Maryland, USA

Tanmay Nath
Department of Biostatistics, Bloomberg School of Public Health, Johns Hopkins University, Baltimore, Maryland, USA, E-mail: tnath3@jhu.edu

Gopal Krishna Nayak
International Institute of Information Technology, Bhubaneswar, Odisha–751003, India

Jipsa Philip
Rajiv Gandhi Institute of Technology, Kottayam, Kerala, India

T. M. Rajesh
Department of Computer Science, Dayananda Sagar University, Bangalore, Karnataka, India

Y. Padma Sai
Department of Electronics and Communication Engineering, VNR VJIET Hyderabad, Telangana, India

Jignyasa Sanghavi
Department of Computer Science and Engineering, Shri Ramdeobaba College of Engineering and Management, Nagpur, Maharashtra, India, E-mail: sanghavijb1@rknec.edu

Sanjay Saxena
International Institute of Information Technology, Bhubaneswar, Odisha–751003, India

S. G. Shaila
Department of Computer Science, Dayananda Sagar University, Bangalore, Karnataka, India, E-mail: shaila-cse@dsu.edu.in

Ravi Shankar Singh
Department of Computer Science and Engineering, IIT (BHU), Varanasi, Uttar Pradesh, India, E-mail: ravi.cse@iitbhu.ac.in

T. Venkata Sridhar
Department of Electronics and Telecommunication, IIIT Bhubaneswar, Odisha, India

Rajeev Srivastava
Department of Computer Science and Engineering, IIT (BHU), Varanasi, Uttar Pradesh, India

Subodh Srivastava
Department of Electronics and Communication Engineering, National Institute of Technology Patna, Bihar, India

Pulkit Thakar
International Institute of Information Technology, Bhubaneswar, Odisha, India

Snekha Thakran
Center for Biomedical Engineering, Indian Institute of Technology Delhi, New Delhi, India, E-mail: snekhathakran@gmail.com

ABBREVIATIONS

ABIDE	autism brain imaging data exchange
AD	axial diffusivity
ADC	analog to digital converter
AFNI	analysis of functional neuroimages
AI	artificial intelligence
AIRNet	affine image registration network
ALU	arithmetic logic unit
ANN	artificial neural network
ApEn	approximate entropy
AUTOMAP	automated transform by manifold approximation
BFC	Bayesian fuzzy clustering
BIDS	brain imaging data structure
BLAS	basic linear algebra subroutines
BOLD	blood oxygen level-dependent
BPNN	back propagation neural network
BSE	blob structure enhancement
CAD	computer-aided diagnosis
CAM	central adaptive medialness
CBIR	content-based image retrieval
CCSDS	consultative committee for space data systems
CEST	chemical exchange saturation transfer
CHESS	chemical-shift selective
CIA	cancer imaging archive
CNN	convolution neural network
CO_2	carbon dioxide
COMA	cache only memory access
CPU	central processing unit
CT	computed tomography
CUDA	computed unified device architecture
DBN	deep belief network
DICOM	digital imaging and communications in medicine
DIP	digital image processing
DITNN	deep learning instantaneously trained neural network

DKI	diffusion-kurtosis-imaging
DL	deep learning
DNN	deep neural network
DPA	deep prior anatomy
DPN	dual path network
DRN	dilated residual network
DSP	digital signal processing
DT	decision tree
DTI	diffusion tensor imaging
DWI	diffusion-weighted-imaging
EDPSO	enhanced Darwinian particle swarm optimization
EEG	electroencephalography
ELM	extreme learning machines
ELU	exponential linear unit
FA	fractional anisotropy
FCM	fuzzy C-means
FCP	functional connectomes project
FD	fractal dimension
FFNN	feed forward neural network
FID	free-induction decay
FIFO	first in first out
FLAIR	fluid-attenuated-inversion-recovery
FLOPS	floating-point operations per second
fMRI	functional magnetic resonance imaging
FPGAs	field programmable gate arrays
FSL	FMRIB software library
GA	genetic algorithm
GAN	generative adversarial network
GBRBM	gaussian-Bernoulli restricted Boltzmann machine
G-CNNs	group equivariant convolutional neural networks
GPU	graphical processing unit
HCP	human connectome project
HPC	high performance computing
HRA	hard region adaptation
HSG	hysterical pornography
HSV	hue saturation value
IDRI	lung image database consortium and image database resource initiative
IED	improvised explosive devices

iEEG	intracranial electroencephalography
ILSVRC	ImageNet large scale visual recognition competition
IoT	internet of things
IPCT	improved profuse clustering technique
IR	inversion recovery
KNN	k-nearest neighbor
KSVM	kernel-SVM
LA	learning automata
LBP	local binary pattern
LDA	linear discriminant analysis
LE	Lyapunov exponent
LSE	line structure enhancement
MBs	megabytes
MC	marching cubes
MD	mean diffusivity
MEG	magnetoencephalography
MEX	MATLAB executable
MGSA	modified gravitational search algorithm
MIMD	multiple instruction, multiple data
MISD	multiple instruction, single data
ML	machine learning
MPI	message passing interface
MPP	message passing protocol
MRI	magnetic resonance imaging
MSD	mean square distance
MTC	magnetization transfer constant
NI	network interface
NIFTI	neuroimaging informatics technology initiative
NIN	network in network
NMR	nuclear magnetic resonance
NN	neural networks
NoC	network on chip
NORMA	no remote memory access
NUMA	non-uniform memory access
ODNN	optimal deep neural network
OS	operating system
PACS	picture archiving and communication systems
PCA	principal component analysis
PD	proton density

PDF	probability distribution function
PE	processing elements
PEIDM	parallel environment for image data mining
PET	positron emission tomography
PSMNET	pyramid stereo matching network
PSO	particle swarm optimization
PVM	parallel virtual machine
QoS	quality of service
RBM	restricted Boltzmann machine
RD	radial diffusivity
ReLU	rectilinear unit
ResNet	residual learning network
RF	radio frequency
RF	random forest
RiR	ResNet in ResNet
RMS	root mean square
RoI	region of interest
RoR	ResNet of ResNet
RRT	recursive ray tracing
RTOS	real time operating systems
SAR	synthetic aperture radar
SCA	sine cosine algorithm
SIMD	single instruction, multiple data
SISD	single instruction, single data
SMs	streaming microprocessors
SNR	signal-to-noise ratio
SoC	system on chip
SPECT	single-photon emission computed tomography
SPM	statistical parametric mapping
SR-FCM	super resolution fuzzy-C-means
SSAC	semi-supervised adversarial classification
SSD	sum of squared distance
STN	spatial transformer network
SVM	support vector machine
TPU	tensor processing unit
UMA	uniform memory access
VPMs	virtual-private-machines
WBC	white blood cells

FOREWORD

High-performance computing has become an essential part of today's mainstream computing systems. In the field of image processing, it has become an essential pillar. For example, point-to-point processing of a grayscale image of dimension 1024 × 1024 necessitates a CPU (central processing unit) to make more than 1 million operations; for color image processing, it is reproduced by several channels. If we consider medical imaging, it has a significant role. During the last few years, the volume of medical image data increased significantly, i.e., from kilo to terabyte. This is primarily due to augmentations in medical imaging acquisition systems with growing pixel resolution and quicker reconstruction processing. For example, the new sky scan 2011 x-ray nano tomography has a 200 nm per pixel resolution and high-resolution micro compound tomography (CT) reconstruct images with 8000 × 8000 pixels per slice with 0.7 μm isotropic detail detectability. The outcome is 64 MB per slice. That is why it is challenging to process such vast amounts of data in a short time. Apart from a massive amount of medical data, other image analysis techniques, specifically deep learning (DL), require high-performing CPUs and GPUs.

Over the past two decades, several research types have been carried out till now, and still, many more are going on to solve this major issue. For example, the efficiency of static and dynamic load scheduling operations on medical image processing algorithms are investigated by several researchers on a multicore environment using several core processors to get the respectable speed up and efficiency with suitable multithreading.

This book provides a valuable window into different medical imaging modalities, different medical image processing techniques, parallel computing, and the need to embed data and task parallelism in medical image processing. The challenges in handling a massive amount of medical imaging data are vast and exciting. Scientists and researchers are working on them with enthusiasm, tenacity, and dedication to develop new parallelism methods, analysis, and provide new solutions to keep up with the ever-changing threats. In this new age of healthcare engineering, it is necessary to provide a way to discuss several issues of medical imaging processing and analysis techniques by both professionals and students with state-of-the-art knowledge on the frontiers in medical sciences. This book is a good step in that direction.

My best wishes are with Dr. Sanjay Saxena and Dr. Sudip Paul (Editors) for their book's success. They have thought of an excellent idea for a book containing information about embedding parallelism in medical image processing.

—Prof. Rajeev Srivastava
Professor and Head of the Department,
Department of Computer Science and Engineering,
Indian Institute of Technology (BHU), Varanasi, Uttar Pradesh, India

ACKNOWLEDGMENT

As nothing can be accomplished by oneself, this work is also not an exception. Though only our name appears on the cover, a great many people have contributed to this work. We would like to have some space to acknowledge them that frequently fade into the background. We want to acknowledge the help of all the people involved in this project and, more specifically, the authors and reviewers that took part in the review process. Without their support, this book would not have become a reality.

First, the editors would like to thank each author for their contribution. My sincere gratitude goes to the chapters' authors who contributed their time and expertise to this book. Secondly, the editors wish to acknowledge the valuable contributions of the reviewers regarding the improvement of quality, coherence, and content presentation of chapters. Most of the authors also served as referees; we highly appreciate their double task.

Further, we want to express our deepest gratitude to Prof. (Dr.) Gopal Krishna Nayak, Director, IIIT Bhubaneswar, India, and Prof. S. K. Srivastava, Honorable Vice-chancellor, North-Eastern Hill University, us the necessary facility to complete this project. We are also thankful to all friends and colleagues whose significant contributions were evident in this book publication in accumulating information and discussing various issues related to the subject under study. They have been very supportive and have encouraged us through every step.

We also want to mention our entire family for their immense support, eternal love, and prayers. This work is possible because of their all-time moral support. We will be failing in our duty if we do not express our thanks to our beloved ones for their continuous support and encouragement.

Last but not least, all praises to Almighty God, who is the source of wisdom.

We thank Him for eternal love and for giving us this opportunity to edit this book.

PREFACE

The main intention of medical image processing techniques is to enhance the quality of medical images and consequently to accomplish feature extraction and classification tasks. Quick handling of imaging data is a noteworthy requirement of several image processing applications, especially in the case of handling diagnostic imaging or medical imaging. In the present day scenario real time, image processing techniques need a massive amount of processing supremacy, computing expertise, and gigantic resources to perform the operations on medical images. The restrictions appear on this system due to the dimension, nature of the image to be a processed and severe increment of data from kilo-to terabyte in the last small number of years. This is mostly due to progressions in medical imaging acquisition systems with growing pixel resolution and quicker reconstruction techniques/algorithms.

High-performance computing (HPC) or parallel computing is the effective and solitary solution to handle this issue, i.e., to obtain high speed of maneuver at real-time resources constraints and full utilization of available resources. In the modern scenario, high-performance computing is the major requirement of enormous and vast computing applications. For this, parallel or distributed computing has become the standard. Parallel image processing delivers a competent approach to handling imaging techniques as the processing of the image makes it clearer, visible, and make suitable for image classification; however, parallelizing of the algorithms elevates the speed at which the image is processed. To achieve the efficient and fastest result in image processing, it is required that the image and its database should be handled in a parallel way using appropriate artificial intelligence (AI) techniques with suitable parallel architecture or environment. This book presents an overview of medical imaging modalities, their processing, high-performance computing, and the need for embedding parallelism in medical image processing techniques.

This book is organized into 12 chapters; it comprises essential chapters written by researchers from prestigious laboratories/educational institutions. A brief description of each of the chapters in this section is given below:

In Chapter 1, different medical imaging modalities such as X-ray, ultrasound, and others with a brief description have been presented. Basically, the origination of vital natural development such as X-ray, ultrasound, radioactivity, magnetic resonance, and the advancement of imaging machinery that handles them have provided some of the adequate diagnostic tools in medicine. With the help of Medical Imaging, people are now capable of investigating the function, pathology, and structure of the human body with a variety of imaging systems. These systems can also be used for arranging treatment or surgery, not only this but also used for imaging in the field of biology. The dataset in 2D, 3D, or high dimensions brings detailed information for research or clinical applications. To be finite health care, this information should be understood in a timely manner. The test is qualitative in some cases, quantitative in other cases; few images are to be registered with each other or with the templates, most of them must be compressed and archived. The international imaging people have matured numerous automated techniques to assist the visual interpretation of medical images, which have their advantages, limitations, and field of application. This chapter presents concepts of various medical image modalities. It is formulated into three sections that resemble the introduction to the medical image modalities, various medical imaging techniques, and conclusion.

Chapter 2 briefs about parallel computing, parallel programming models, and multicore architectures. In this current age of computation and networking, the processor always holds the top priority of any kind of computation. With the recent increase in the diverse fields of data, special visual data requires high-performance computing (HPC). To facilitate the HPC, parallel computing became an instrumental science at its inception. Parallel computing is now one of those fields that are being researched very vigorously and enthusiastically. The parallel computing machine is the massively parallel architecture of processors that provides real-time and simultaneous processing of high-end tasks. In this study, we provide all aspects of parallel computing begins with the overall architecture details, memory architecture in parallel computing, which includes shared memory, distributed memory, and hybrid memory. Flynn's classification of computer architecture and von Neumann architecture always holds a special place in parallel computing are also illustrated with the theme of parallel processing. The graphical processing unit (GPU) with CUDA processing, nowadays became trending in the study of parallel processing, are highlighted in this chapter. Last but not least, the embedding of parallelism in the study of image processing with real-time case studies is explored. So, this chapter will enlighten and strengthen all the requirements regarding parallel computing of the potential researcher for their further research in this domain.

Chapter 3 provides a brief description of the fundamentals of medical image processing. In 1895 after the discovery of x-ray, images were commonly used for medical diagnostics. The process of taking images of the body parts for medical uses in order to study or identify various diseases is known as medical imaging. Throw out the world every week there are millions of imaging procedures been done. In this rapidly growing medical imaging process, including image recognition, analysis, and enhancement in image processing techniques. Image processing increases the percentage and amount of detected tissues. In this chapter, the authors presented a basic understanding of medical images and processing. Image processing is often viewed as arbitrarily manipulating an image for achieving an esthetic standard or supporting a preferred reality. Why we do the image process is because the fact is that the human visual system does not recognize the world in a similar way as sensors that record in the form of digital data.

Chapter 4 briefs about multicore architectures which are the widely popular design techniques that are used in most electronic devices nowadays. Having many cores give a quicker and timely accurate response in many applications, especially like medical image processing, which is a more critical and essential area. Integrating many cores into a single chip is a big task as both software, hardware, and firmware must coordinate simultaneously for better yield. Here in this chapter, the authors tried to cover the basics concepts to moderately high levels of abstraction to understand the three coordination (Hardware vs. Software vs. Firmware). To produce accurate and good results, it is required to understand the compatibilities and issues in combining the hardware and software co-design. Because the field of medical image processing critically requires the rate of false-positive as minimum as possible as it deals with lives.

Chapter 5 briefs about the term learning relate to a broad range of processes that can be defined as "knowledge or skill acquired by instruction or study." Psychologists study learning in both humans and animals, but in this chapter, the authors focused on learning in machines and how this process has evolved during the last 60 years and found important applications in medical image processing. The author discussed how computational models built to understand human learning have led to two main categories of machine learning: supervised learning and unsupervised learning. They also presented the concepts of deep learning and how it is beginning to play a vital role in medical image processing. To help researchers interested in applying machine learning to medical imaging data, they have provided information about different data formats and available resources to analyze these data. The chapter concludes with a discussion of how machine learning is expected to continue to play an important role in medical image processing and, combined with a doctor's experience, will help improve medical outcomes.

Chapter 6 briefs about magnetic resonance imaging (MRI), also known as nuclear magnetic resonance imaging, is a special acquisition technique that creates minute images of the body. The scan uses a magnetic field (weak/medium or strong: 0.2, 1.5, 2, 7T) and radio waves to create images of parts of the body. These parts are otherwise not visible well with X-rays, CT scans, or ultrasound. One such example would be: it can help clinicians to see joints inside the body, ligaments, cartilage, muscles, etc. This is useful for detecting various sports injuries. MRI helps to diagnose multiple disorders and can examine the internal body, such as spinal cord injuries, strokes, aneurysms, multiple sclerosis, tumors, and eye or inner ear problems. It is also widely used in research to measure brain structure and function, in addition to multiple other things. An MRI is an extremely accurate method of disease understanding. After the other testing methods fail to provide sufficient enough information to confirm a patient's diagnosis, MRI is generally the last resort. In the head, bleeding/swelling can indicate brain trauma. Several other abnormalities are also detected via MRI, and some examples would be: aneurysms, stroke, tumors of the brain, tumors, and inflammation of the spine. MRI scans help neurosurgeons to visualize the brain anatomy, spinal cord integrity in trauma situations, diseases associated with the spine, such as vertebrae or intervertebral disc issues. In heart and aorta structures, in detecting aneurysms or tears, MRI proves to be a valuable component. Especially in glands and organs of the abdomen, MRI can provide valuable information. In the case of joints structures, soft tissues, and bones of the body, a high-resolution MRI can provide minute details that are otherwise not visible in any other modality. MRI scans help to confirm the surgical outputs, progress tracking, treatment planning, and several such cases. In this chapter, the authors summarized conventional and advance magnetic resonance imaging methods.

Chapter 7 briefs about brain tumors are the phenomenal growth of abnormal tissues in the human brain. A malignant tumor is known as a cancerous cell, and it is the major cause of death among people across the globe. Nowadays, the detection and classification of brain tumors from Magnetic Resonance Images (MRI) is a very crucial task. This chapter addresses various computing methods such as Edge Detection (ED), feature extraction, and classification techniques for the detection and classification of brain tumor regions from MRI datasets. Initially, MRI images are collected, and then various pre-processing steps such as filtering and edge detection are applied. Then, different edge detection methods such as Sobel ED, Robert's ED, Prewitt ED, Canny ED, Laplacian ED, and Laplacian of Gaussian (LoG) with sigma 3 are applied for the MRI datasets. Subsequently, segmentation techniques are applied for detecting tumor regions from MRI, and essential features are extracted using Discrete Wavelet Transform (DWT) method. Otsu's thresholding and K-means clustering segmentation methods are used for the investigation. Further, the support vector machines (SVMs), Naïve-Bayes (NB), K-nearest neighbors (KNN), back propagation neural network (BPNN), and feedforward neural networks (FFNN) classifiers are employed for the MRI classification purpose. The experiments are conducted using the brain tumor dataset in the MATLAB 2019a software environment. The experimental results are analyzed in multiple dimensions, and it shows that the SVM with Otsu's thresholding method exhibits better performance with 86.11% accuracy during the classification.

Chapter 8 addressed two complex issues segmenting 2D slices of CT or MRI and view segmented 2D slices in 3D viewer. For segmenting 2D slices, we used the watershed algorithm. The 3D viewer is used for better visualization, which gives more information about segmented tumors. For better segmentation, included Gradient technique with the watershed algorithm. Segmented series of 2D images are viewed in 3D viewer. Experimental result shows viewing segmented 2D slices in 3D viewer. In our proposed methodology, we have shown the essence and feasibility of an automated tumor segmentation method for both CT and MRI images, and a simple model consists of the watershed algorithm is used to segment the tumor. The method segments a series of slices that consists of tumors and can be viewed using the 3D viewer technique and has been validated on five clinical MRI datasets, which consists of a minimum of 20–25 slices. The end results show a promising result.

Chapter 9 briefs about the recent advancements in the field of computer science that have led to the development of systems with the massive computing power of 1015 floating-point operations per second. High-performance computing (HPC) systems use a large number of computing nodes to attain that performance practically. Parallel computing is a branch of HPC that focuses on reducing the execution time of any application. This work focuses on embedding parallelism in three primary hyperspectral image (HSI) processing techniques that are widely used in many applications like object identification, military operations, food security, monitoring natural disasters, and many more. Parallelism in these techniques is required as they work on complex mathematical operations and learning algorithms having high runtime. Some applications of these techniques work on real-time processing having energy and time constraint. It provides a comparative analysis of some recent and trending research works in the field of HSI segmentation, HSI compression, and HSI classification. Important evaluation metrics for parallel algorithms used in subsequent works have been described in detail. The purpose of this chapter is to provide a survey along with future research directions in the field of parallel image processing techniques to get the benefit of parallel computing in HSI processing.

Chapter 10 briefs according to World Health Organization, cancer is the second leading cause of death globally, and is responsible for an estimated 9.6 million deaths in 2018. Globally, about 1 in 6 deaths is due to cancer. Among cancers, lung cancer is the leading cause of death. Lung cancer is the abnormal and uncontrolled growth of cancerous cells in the lungs. These cells destroy the nearby tissues, and cells form cancerous nodules. Automatic computerized detection systems are used for the early detection of lung cancer. In this chapter, the systematic review of automatic detection of the pulmonary nodule from chest CT scans using image processing and machine learning techniques is illustrated. The automatic CAD (computer-aided diagnosis) systems from simple CBIR systems using similarity functions to complex deep neural networks with false-positive reduction techniques are summarized in this chapter. This chapter gives a brief introduction of methods and techniques used in the processing stages of an automatic system like pre-processing, segmentation, feature extraction, feature optimization, and classification.

Chapter 11 briefs about image processing become demanding and attracting research domain due to its versatile application in a different field such as military field, medical imaging, authentication of the digitization process through signature recognition, face recognition, the agricultural field, etc. Due to improved technology and massive development in cost-effective image acquisition equipment, the size and number of images increase day-by-day. In some domains, such as medical, military applications, accuracy, as well as real-time image analysis, is the most important criterion for evaluating the performance of the system. As each application is having its own requirement, every system demands real-time, accurate, less expensive, and more extensive computation. Most of the application of computer vision, image processing, pattern recognition deploys the feature extraction to use the machine learning and deep learning methodology. All these techniques and architecture requires huge time to extract the features of the images which could not be acceptable in some of the real-world application. To reduce the time and make the computation faster and efficient, the concept of parallel processing is embedded in image analysis. This study aims to provide an introduction as well as the need of parallelism in image processing. The mostly used parallel processing using CUDA and GPU is briefly discussed with their shortcomings. Finally, how parallel computation is used in machine learning and deep learning architecture for medical image analysis is discussed in detail to show the impact of parallel processing. Also, one case study based on brain tumor segmentation is elaborated in this chapter. The main purpose of this chapter is to summarize existing parallel image processing techniques and tools via various research and analysis and their limitation.

Chapter 12 briefs about modern-day computing involve computation of complex, multi-dimensional, and volumetric data. The generation of these data, basically from the web world, networking, and high-end applications. Natural images, medical images, hyperspectral images, and video datasets are always in complex form. To handle these data and run the application that generating these kinds of data and especially solve the complex problems related to these dataset, high-performance computing is the ultimate choice. With the recent success and trending of deep neural networks for solving computer vision applications is a big leap in the computation direction. So to solve complex computer vision problems with deep neural networks, high-performance computing is again the good choice, as deep neural networks are with deep architecture and many more trainable parameters. In this chapter, the authors cover many more aspects of high-performance computing and its requirements in deep learning in a competitive approach to help the potential research further.

The objective of this book is to provide information on medical image processing techniques, parallel computing techniques, and the need for embedding parallelism in different image processing techniques. Further, this book will provide relevant theoretical frameworks and the latest empirical research findings in the area. The comprehensive review of parallel algorithms in medical image processing problems will be the key feature of this book.

The target audience of this book will be composed of professionals, researchers, and students working in the field of health care engineering, medical imaging technology, its handling, applications of machine and deep learning. Further, this book may be recommended as the curriculum of graduate students, postgraduate students in computer engineering, biomedical engineering, and electrical engineering based on artificial intelligence, parallel computing, high-performance computing, machine learning, and its applications in medical imaging.

The editors wish you a pleasant reading.

—**Sanjay Saxena, PhD**
Sudip Paul, PhD

CHAPTER 1

BASIC UNDERSTANDING OF MEDICAL IMAGING MODALITIES

PRADEEP KUMAR,[1] SUBODH SRIVASTAVA,[1] and
RAJEEV SRIVASTAVA[2]

[1]*Department of Electronics and Communication Engineering,
National Institute of Technology Patna, Bihar, India*

[2]*Department of Computer Science and Engineering, IIT (BHU),
Varanasi, Uttar Pradesh, India*

ABSTRACT

The origination of vital natural development such as X-ray, ultrasound, radioactivity, magnetic resonance, and the advancement of imaging machinery that handles them have provided some of the adequate diagnostic tools in medicine. With the help of medical imaging, people are now capable of investigating the function, pathology, and structure of the human body with a variety of imaging systems. These systems can also be used for arranging treatment or surgery, not only this but also used for imaging in the field of biology. The dataset in 2D, 3D, or high dimensions brings detailed information for research or clinical applications. To be finite health care, this information should be understood in a timely manner. The test is qualitative in some cases, quantitative in other cases; few images are to be registered with each other or with the templates, most of them must be compressed and archived. The international imaging people have matured numerous automated techniques to assist the visual interpretation of medical images, which have their advantages, limitations, and field of application. This chapter presents concepts of various medical image modalities. It is formulated into three sections that resemble the introduction to the medical image modalities, various medical imaging techniques, and conclusion.

1.1   INTRODUCTION

Visualization plays a vital role in medical imaging applications. With the discovery of various medical imaging techniques such as x-rays, ultrasound, radioactivity, magnetic resonance, and the advancement of imaging machineries that tackle them have provided some of the most adequate diagnostic apparatus in medicine. Because of above mentioned techniques of medical imaging visualization is possible in biomedical applications [1, 2]. The imaging process used in biology and medicine are established on a various energy source, including light, electrons, x-rays, radionuclide's, lasers, Ultrasound, and nuclear magnetic resonance (NMR). Magnitude in scale is created by producing images of span ordered, this scale is raining from molecules, and cells to organ systems, and full body [3, 4]. The merits and demerits of each imaging process are mainly controlled by the primary biological and physical principles which influence the way each energy form interacts with tissues, and by the explicit engineering implementation for a particular medical or biological application. the various illness process and abnormalities affecting all parts of the human body are so high and distinct that each imaging process have characteristics that make it distinctly helpful in furnishing the desired understanding or prejudice of the illness or abnormality, and so, no single procedure has assured to the complete boycott of others. Even though noteworthy distinction in scale or attribute features isolate the imaging fields, striking parallel and common approach prevail relative to visualization and investigation of these images [1]. In general, the techniques are complementary, together providing a powerful and synergistic armamentarium of clinical diagnostic, therapeutic, and biomedical research capabilities which has potential to significantly leading the practice of medicine.

1.2   TYPES OF MEDICAL IMAGE MODALITIES [5]

- Conventional radiography (X-ray);
- Fluoroscopy;
- Angiography;

- Mammography;
- Computer tomography (CT);
- Ultrasound and ultrasound/doppler;
- Magnetic resonance imaging (MRI);
- Tomosynthesis;
- Nuclear medicine;
- Positron emission tomography (PET);
- Positron emission tomography/computed tomography (PET/CT);
- Thermal imaging;
- Micro biopsy.

1.2.1 CONVENTIONAL RADIOGRAPHY

Conventional radiography is a technique in which x-rays are used to visualize the internal structure of patient, intermittently used imaging types is an x-ray; it is the earliest imaging process. We all know and have at least been taken an x-ray during our course of live. This is a form of electromagnetic radiation and was discovered in 1985 [3]. We are unable to see the x-rays with our eyes. The detector captures the x-rays crossing the body of the patient; films are sensitive to x-rays or a digital detector. The absorption of x-rays within the body varies from tissues to tissues; dense bone consumes more radiation, while the soft tissue grants more to pass through. The 2D image of all the structures within the patient is formed by the deviation produce when x-rays beam is passed. X-rays are typically used for analyzing issues with chest, skeleton, abdomen, and dental. Much radiation of x-rays is lethal for patient. So, the x-ray personnel should confirm with patient is naked to as little radiation as possible (Figure 1.1) [5].

1.2.2 FLUOROSCOPY

Fluoroscopy is an imaging process which has great real-time visualization of body structures by using x-rays. At the time of fluoroscopy, a real-time dynamic image is generated by emitting x-ray beams constantly and then is captured on a screen [5]. To allow high distinction between structures, a high anatomy density contrast agent will be launched into the patient; this allows the dynamic evaluation of function. Fluoroscopy is typically used for recognizing issues with barium studies, hysterical pornography (HSG), histography, and reduction of fractures under image guidance. Radiation doses from fluoroscopy are higher because of that pregnancy status must be known before the test. Children are more tactful to radiation so x-ray procedure should be performed with caution (Figure 1.2).

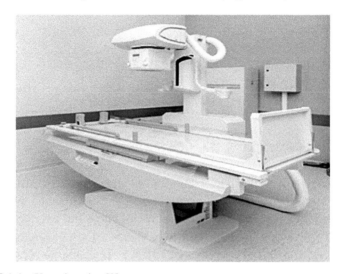

FIGURE 1.1   X-ray imaging [2].

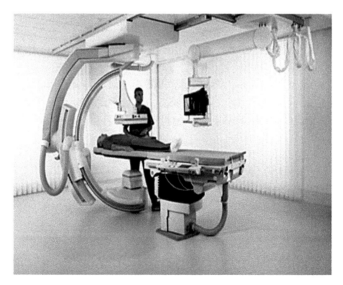

FIGURE 1.2   Fluoroscopy imaging.

Source: Image courtesy of Siemens Healthcare, USA.

1.2.3 ANGIOGRAPHY

Angiography is a type of medical image process which is used for the anticipation of the inside of blood vessels, especially the arteries, veins, and the heart chambers. For the study of angiography contrast media is infused into the blood vessels [8]. In this method also we use x-rays. There are two types of angiography machines, conventional angiography, or digital subtraction angiography. Commonly used angiography is digital subtraction angiography. Presently, most of the measures achieved by angiography are replaced by CT angiography or magnetic resonance angiography. Angiography is typically used for diagnosing issues with obstructive vascular diseases, bleeding vessels, arterio-venous malformations, and assessment of the vascularity of malignant tumors. Radiation doses from fluoroscopy are higher because of that pregnancy status must be known prior to test. Children are more tactful to radiation so x-ray procedure should be performed with caution (Figure 1.3) [5].

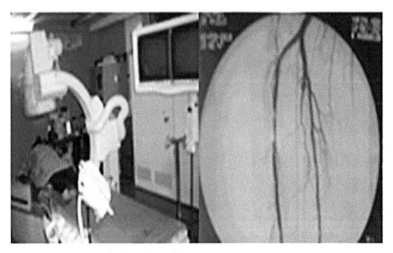

FIGURE 1.3 Angiography imaging [5].

1.2.4 MAMMOGRAPHY

Mammography is an imaging modality used for imaging of breast tissue. Mammography uses low energy x-rays. Mammography exercise exploits systematize views of the breasts to the estimation of breast lesions [15–17].

In asymptomatic women early detection of breast cancer are done by mammography. For ultimate visualization of masses or calcifications the breast is tested isolated and squeezed one by one in between the film. The early treatment and increased rates of survival can be done by the help of early detection of breast cancer. Typical clinical applications of mammography are screening mammography, diagnostic mammography, surveillance mammography and needle localization and tumor marking [11]. Mammography is a safe procedure because of low energy x-rays. During examination, pain may occur sometimes because of the squeezing of the breast tissue [5]. Bleeding also can occur in some patients due to biopsy (Figure 1.4).

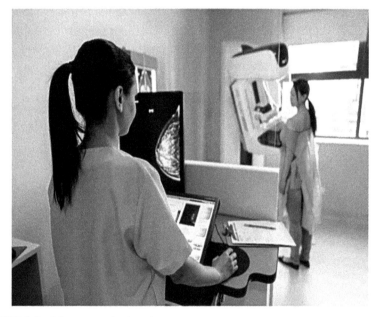

FIGURE 1.4   Mammography imaging.
Source: www.hopkinsmedicin.org/.

1.2.5 COMPUTER TOMOGRAPHY

Computer Tomography (CT) is a type of image processing that creates a 3D picture for diagnosis using x-ray photons. The main parts of the CT scanners are detectors and x-ray tube. The 2D or 3D images are created by reconstruction of beam captured by the detectors released from x-ray tube which crosses over the patient. As compare to conventional X-rays CT scans provide better clarity [3]. The data captured by CT scanner is in analog form and is converted into digital with the help of numerous algorithms. At that level it represents a cross-sectional slice through the patient. Different reconstruction algorithm is required for slightly different angle of image acquired from results. In CT scan study, contrast media may be employed to distinguish structures of similar density in the body. Many irregularities such as extravasation, bleeding, or neoplasms become more visible through contrast perfusion. Typical clinical applications of CT scanners are screening brain, CT myelography, Cardiac CT and quantitative computed tomography [5]. X-rays are been used by CT scan to generate images: anyway, because of multiple exposures the radiation comes from CT scan are higher than plain radiography. The status of pregnancy must be confirmed prior to test. At time of children extra caution should be taken because children are more radiosensitive (Figure 1.5).

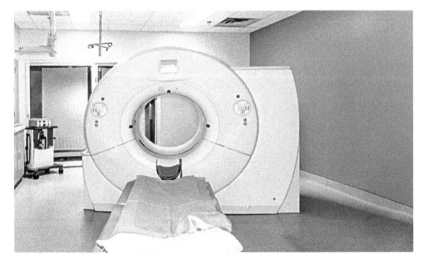

FIGURE 1.5   CT scan imaging [2].

1.2.6   ULTRASOUND AND ULTRASOUND/DOPPLER

Ultrasound uses sound waves of high frequency to provide cross-sectional images of body. This form of medical imaging is very safe. There are a wide range of applications. The ultrasound machine has the following components: keyboard, monitor, data storage processor, and transducer. At a certain frequency the transducer emits echoes; those echoes are captured at frequencies dependent on tissues through which the waves traverse [3]. The echoes or dots appearing on screen are the sound waves the returns to transducer. "Doppler" is another type of ultrasound commonly used; it has a little difference in using sound waves that allows the vascular studies. It is less in cost and easy to perform, easy to move to other places. The result is operator dependent. The main clinic application of ultrasound is in pregnancy because of low risk. The other clinical applications are abdominal, pelvic, obstetric, cardiovascular ultrasound including echocardiography and trans fontanelle ultrasound. With no radiation and minimal adverse effects, it is known as very safe procedure and also widely used in antenatal care (Figure 1.6) [5].

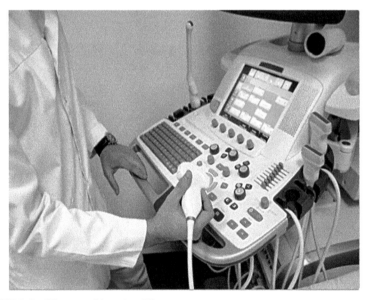

FIGURE 1.6   Ultrasound imaging [2].

1.2.7   MAGNETIC RESONANCE IMAGING (MRI)

Magnetic resonance imaging (MRI) is a type of image modality which uses magnetic radiation to visualize complete internal structure. The real time 3D view of organs with good soft tissue contrast and creating a view of brain, spine, muscles, joints, and other structures excellent [5]. Getting multiple body planes images without changing the position is known as multiplanar image. Before going to MRI examinations, the patient is said to remove all metallic objects which includes watches, jewelry, and piercings. Ear plugs are usually provided because of loud noises. Typically, MRI is used for the diagnosis of brain, breast, chest/ mediastinal MRI, cardiac, and image-guided interventional procedures. There is no radiation exposure in it which makes MRI safe [6]. A patient with metallic implants such as pacemakers, surgical clips or other foreign bodies can cause death or have injuries while MRI. For such patient's extreme caution must be taken to avoid MRI. There is no side effect for pregnant women (Figure 1.7).

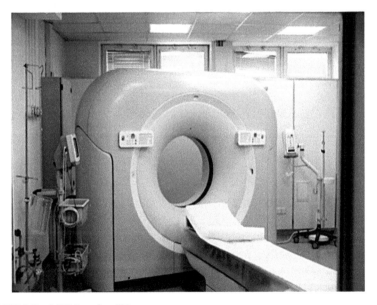

FIGURE 1.7   MRI imaging [2].

1.2.8 TOMOSYNTHESIS

Tomosynthesis is a type of imaging modality which can be used to screen for early signs of breast cancer in women with no symptoms. This can also be used as the diagnostic tool for women having these breast cancer symptoms. It is an advanced version of mammography. Multiple breast images are taken by tomosynthesis. An algorithm to combine them into a 3-D image of the entire breast is done by the images sent to the computer. This process is mainly used for breast cancer. It is also using radiation like x-ray so we should be conscious at the time of using tomosynthesis (Figure 1.8) [9].

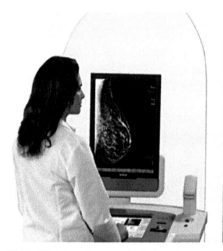

FIGURE 1.8   Tomosynthesis imaging.
Source: www.breastlink.com.

1.2.9 NUCLEAR MEDICINE

Nuclear Medicine is an imaging procedure which includes inhalation, injection, or radioactive traces injection to observe various organs. An organ to be captured it uses a tracer or radiopharmaceutical which implemented with addition of a radioactive isotope to a pharmaceutical specific to the organ. Using a gamma camera is further used to form image, this gamma radiation is emitted from radioactive tracer. The sensitive radiation crystal in gamma camera is used to detect the tracer distribution in the patient's body. The data from it is converted to digital form to generate 2D or 3D image on screen. The hybrid machine including a CT to allow the fusion of Nuclear Medicine and CT images come under the latest technology of gamma camera [8]. Therapy procedures are also indulging in field of Nuclear Medicine. Diseased target organ is identified by giving high dose of therapeutic radiation through administration of radiopharmaceutical specific to organ. To treat cancer or over functioning thyroid gland it is beneficial. Typical clinical applications of nuclear medicines are bone scan, myocardial Perfusion scan, renal scan, lung scan and thyroid scan. Pregnancy status must be done before the procedure. Patients who are undergoing therapy procedures will be given specific set of instructions regarding radiation safety (Figure 1.9) [5].

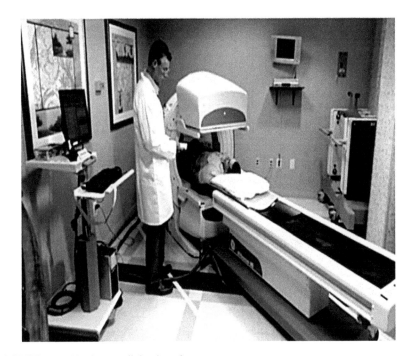

FIGURE 1.9 Nuclear medicine imaging.
Source: www.radiologyinfo.org.

1.2.10 POSITRON EMISSION TOMOGRAPHY (PET)

Positron emission tomography (PET) is a type of image modality which is used to study about how your tissues and organs are functioning. It is a type of nuclear medicine [12]. A radioactive drug (tracer) is used in PET scan to show above mention activity. This scan can sometimes disclose disease before it shows up on other imaging tests. The tracer may be injected, swallowed, or inhaled, depending upon which organ or tissue is being tested. A PET scan can be used for several cancers, heart diseases and brain disorders. It is mostly used for chemical activity. PET scans must be understood attentively because sometimes noncancerous disease looks like cancerous and sometimes cancers do not appear also (Figure 1.10).

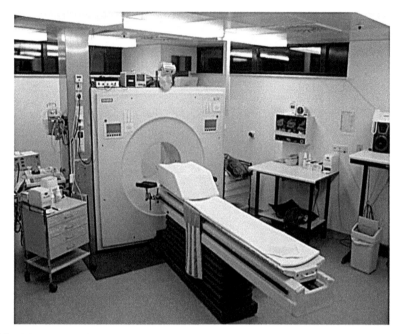

FIGURE 1.10   Positron emission tomography imaging.

Source: https://en.wikipedia.org/wiki/Positron_emission_tomography.

1.2.11   POSITRON EMISSION TOMOGRAPHY/COMPUTED TOMOGRAPHY (PET/CT)

Many test centers superimpose nuclear medicine image with CT to produce special views. This state is known as image fusion [8]. This view allows doctor to correlate and interpret two different tests on one image which leads to more precise and accurate information. PET/CT has reformed medical investigation in many fields, by adding precision of anatomic localization to functional imaging, which was previously lacking from pure PET imaging. For most of the diagnosis such as oncology, surgical planning, radiation therapy and cancer staging, PET/CT is used at the place of PET. Now a days every one changing from PET to PET/CT. The most difficulty in using PET/CT is its cost (Figure 1.11) [10].

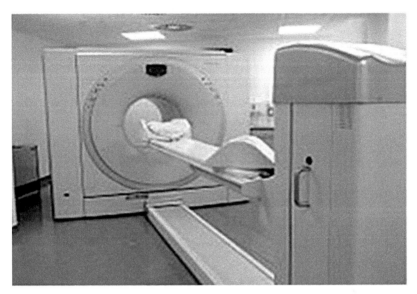

FIGURE 1.11   PET/CT imaging (siemens biograph PET-CT scanner).

1.2.12   THERMAL IMAGING

Thermal imaging is a type of image modality which is used to convert infrared radiation into detectable images that represent the spatial distribution of temperature differences in a scene explored by a thermal camera. In thermal imaging, an infrared detector is equipped with camera, mostly in a focal plane array, of micron-size detecting elements or "pixels" [13]. Depending on the materials comprising the array and the camera's intended use, the detector array may be cooled or uncooled. It is important to generate the finest images conceivable to extract purposeful input regarding the detection, recognition, and identification of animals of interest in the field. This is absolutely the ambition of surveillance applications, which the military has been laboring over for years (Figure 1.12).

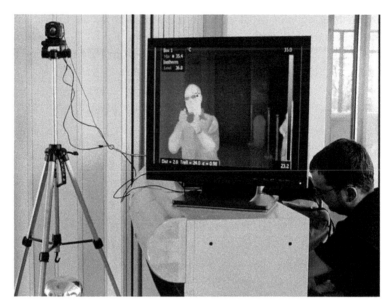

FIGURE 1.12   Thermal imaging.

Source: https://en.wikipedia.org/wiki/Thermography.

1.2.13   MICRO BIOPSY

A biopsy is a medical test commonly performed by a surgeon, interventional radiologist, or an interventional cardiologist involving extraction of sample cells or tissues for examination to determine the presence or extent of a disease. The micro biopsy is a relatively new biopsy technique which allows muscle physiologists to sample skeletal muscle less invasively. The sample size is too small. It seems insufficient for certain analysis [14]. Wireless capsule endoscopy is also one type of micro biopsy (Figure 1.13).

1.3   CONCLUSION

Medical imaging modalities provide unique windows into the structure and human body biochemistry. Because of the new imaging models, more similar simulation models, and continued growth in computational power all contribute to challenging biomedical researchers, engineers, doctors, and with an unprecedented volume of information to further their knowledge of biological systems and improve their practice in clinical system. With the help of this modality's technique, it is very easy for doctor to save the people. It is also good in the field of research. We are able to get abundant data sets for research work and able to provide second opinion for doctor. Anyway, Visualization plays a vital role in medical imaging applications. In this chapter, we have provided a selected overview of recent image modalities, while emphasizing promising avenues for future research.

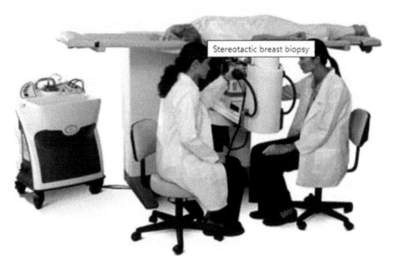

FIGURE 1.13   Patient undergoing stereotactic breast biopsy.

Source: www.radiologyinfo.org.

KEYWORDS

- **computer tomography**
- **hysterical pornography**
- **image processing technique**
- **magnetic resonance imaging**
- **positron emission tomography**
- **visualization**

REFERENCES

1. Richard, A. R., (1999). *Biomedical Imaging, Visualization, and Analysis*. pp. 1–360. ISBN: 978-0-471-28353-9. John Wiley & Sons, Inc.; 1st edition.
2. Yousif, M. Y. A., & Tariq, A., (2019). Research in medical imaging using image processing techniques. In: Yongxia, Z., (ed.), *Medical Imaging - Principles and Applications*. IntechOpen. doi: 10.5772/intechopen.84360.
3. Hughes, Z. (2021). *Medical Imaging Types and Modalities*. https://www.ausmed.com/cpd/articles/medical-imaging-types-and-modalities (accessed on 29 September 2021).
4. Kasban, H., El-Bendary, M., & Salama, D., (2015). A comparative study of medical imaging techniques. *International Journal of Information Science and Intelligent System, 4*, 37–58.
5. World Health Organization, (2020). *Imaging Modalities*. WHO, Geneva.
6. Nordbeck, P., Ertl, G., & Ritter, O., (2015). Magnetic resonance imaging safety in pacemaker and implantable cardioverter defibrillator patients: How far have we come? *European Heart Journal, 36*(24), 1505–1511.
7. Madhuri, A. J., (2010). *Digital Image Processing - An Algorithmic Approach*. Prentice Hall of India.
8. Yang, W., & Liu, J., (2013). Research and development of medical image fusion. In: *2013 IEEE International Conference on Medical Imaging Physics and Engineering* (pp. 307–309). Shenyang.
9. Miroshnychenko, S. I., Nevhasymyy, A. A., Senchurov, S. P., & Motolyga, O. V., (2014). The implementation of the digital tomosynthesis mode into the radiological table modality. In: *2014 IEEE 34th International Scientific Conference on Electronics and Nanotechnology (ELNANO)* (pp. 345–347). Kyiv.
10. Zhong, Z., et al., (2018). Improving tumor co-segmentation on PET-CT images with 3D co-matting. In: *2018 IEEE 15th International Symposium on Biomedical Imaging (ISBI 2018)* (pp. 224–227). Washington, DC.
11. Remeš, V., & Haindl, M., (2015). Classification of breast density in x-ray mammography. In: *2015 International Workshop on Computational Intelligence for Multimedia Understanding (IWCIM)* (pp. 1–5). Prague.
12. Starfield, D. M., Rubin, D. M., & Marwala, T., (2007). High transparency coded apertures in planar nuclear medicine imaging. In: *2007 29th Annual International Conference of the IEEE Engineering in Medicine and Biology Society* (pp. 4468–4471). Lyon.
13. Bozhenko, V., Kondratov, P., & Tebenko, J., (2012). Thermal images expert evaluation. *Proceedings of International Conference on Modern Problem of Radio Engineering, Telecommunications and Computer Science* (p. 156). Lviv-Slavske.
14. Cosnier, M. L., Martin, F., Bouamrani, A., Berger, F., & Caillat, P., (2006). A new micro minimally invasive biopsy tool for molecular analysis. In: *2006 International Conference of the IEEE Engineering in Medicine and Biology Society* (pp. 2820–2823). New York, NY.
15. Srivastava, S., Sharma, N., & Singh, S. K., (2014). Image analysis and understanding techniques for breast cancer detection from digital mammograms. In: *Research Developments in Computer Vision and Image Processing: Methodologies and Applications* (pp. 123–148). IGI Global.

16. Srivastava, S., Sharma, N., Singh, S. K., & Srivastava, R., (2014). Quantitative analysis of a general framework of a CAD tool for breast cancer detection from mammograms. *Journal of Medical Imaging and Health Informatics, 4*(5), 654–674.
17. Srivastava, S., Sharma, N., Singh, S. K., & Srivastava, R., (2013). Design, analysis and classifier evaluation for a CAD tool for breast cancer detection from digital mammograms. *International Journal of Biomedical Engineering and Technology, 13*(3), 270–300.

CHAPTER 2

PARALLEL COMPUTING

BISWAJIT JENA, PULKIT THAKAR, GOPAL KRISHNA NAYAK, and
SANJAY SAXENA

*International Institute of Information Technology, Bhubaneswar,
Odisha, India, E-mail: biswajit310@gmail.com (B. Jena)*

ABSTRACT

In this current age of computation and networking, the processor always
holds the top priority of any kind of computation. With the recent increase
in the diverse fields of data, special visual data requires high-performance
computing (HPC). To facilitate the HPC, parallel computing became an
instrumental science at its inception. Parallel computing is now one of
those fields that are being researched very vigorously and enthusiastically.
The parallel computing machine is the massively parallel architecture
of processors that provides real-time and simultaneous processing of
high-end tasks. In this study, we provide all aspects of parallel computing
begins with the overall architecture details, memory architecture in parallel
computing, which includes shared memory, distributed memory, and
hybrid memory. Flynn's classification of computer architecture and von
Neumann architecture always holds a special place in parallel computing
are also illustrated with the theme of parallel processing. The graphical
processing unit (GPU) with CUDA processing, nowadays became trending
in the study of parallel processing, are highlighted in this chapter. Last but
not least, the embedding of parallelism in the study of image processing
with real-time case studies is explored. So, this chapter will enlighten
and strengthen all the requirements regarding parallel computing of the
potential researcher for their further research in this domain.

2.1   INTRODUCTION

2.1.1   WHAT IS PARALLEL COMPUTING?

Clearly "Parallel Computing" comprises of two words: Parallel and Computing [1]. The term parallel here refers to simultaneously, and the term computing means executing a program. This can give us the idea of what parallel computing means. Parallel computing is now one among those fields that is being researched very vigorously and enthusiastically. There is a fine border between parallel computing and concurrent computing. Concurrent computing refers to "in progress at the same time." And parallel computing refers to simultaneous working of programs. Here the difference is that concurrent computing is usually done for uni-processor machines. And parallel computing is used by multi-processor machines, hence letting many processes to be run simultaneously.

The need of such computations was visible way back when the mainframe machines were used. The processors always were very fast in processing since the very beginning. But, the input and the output of data used to take a lot more time than the processor to process the data. Above this, the machines could only process only one process at a time. This gave birth to the one among the first problems of parallel computations. Here it is visible to us that the processor is idle for all the time when the data is being input and also when the data is being output. The problem was framed-if there could be a way to allow the processor work continuously if the data was readily available- and so the research began. Different methods are being developed to keep the processor busy and all the work that does not need the involvement of processor to be performed separately.

Various and numerous problems were seen. The above was one illustration of them. The research still stands and has become vast. The use of multi-processor machines was a huge breakthrough for these problems. The usage of multi-processor machines is now very common. The desktops now being used at homes are also multi-processor machines. Till the machines were uni-processor the concurrent computing was being used. But once the multi-processor machine came, the concurrent computing in them was nothing but parallel computing. Since then, the terms concurrent computing and parallel computing are being used interchangeably [2–5].

2.1.2   WHY DO WE USE PARALLEL COMPUTING?

Parallel computing enables us to fasten any process. Any task can be divided into sub-tasks. These sub-tasks do not have any relation between them, i.e., that they are independent of each other. This allows them to run separately. In a multi-processor system, each sub-task is assigned to a separate processor and the results of all the sub-tasks are combined together to get the final result of the task. How does this speed up the process? Let us assume that one processor takes up the whole task and runs the code line by line. We will get the result only once the whole program is executed. Now let us assume that the task is divided into two halves and are executed in two different processors. In this case the time taken for each processor to execute its part will be half the time taken to execute the whole task. When both the processors run simultaneously performing their parts, then time taken by both of them will be same, which is half of the time taken to perform the whole task by a processor. This evidently fastens the process. Now we can get an overview of what would be the effect of this if a task is divided into even more number of sub-tasks. Figure 2.1 lets us understand the concept of parallel computing easily. The concept of parallel computing has also been extended to multiple processes. This allows the overall work done by the processor to speed up.

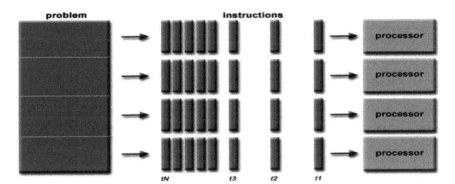

FIGURE 2.1   Architecture of parallel computing.

Source: Reprinted from Ref. [1].

The above gives us an insight about the general meaning of parallel computing. And the need of parallel computing is now being evidently visible in every field which needs high computation. Image processing is one such field. When it comes to image processing in the field of medicine then the images that are to be dealt with might be of dimensions more than 2. An example can be taken of the MRI scans. The MRI scans have images of the brain and in order to cover the whole brain the scans result in 3-dimensional images. Therefore, the use of parallel computing has also been adopted in image processing. This, as expected, will speed up the computation. Is speeding up of the process so essential? This question can be answered by taking the example of real-time image processing, such as live detection of traffic using the data from the CCTV cameras. The outputs for such problems are to be generated in real-time as well. Processing such input data to get the results right away will not be possible by using traditional serial computing. Parallel computing proves its stand here. So, this chapter intends to go through the concepts of parallel computing for a better understanding of it by covering the following: memory architecture used for parallel computing, basics, and terminology in parallel computing, computing models, GPU in parallel computing and some case studies.

The remaining parts of this chapter are as follows: Section 2.2 will discuss about the memory architecture in parallel computing. Next, Section 2.3; explore the basic terminologies used in parallel computing study. Section 2.4, highlights various parallel programming models used in parallel computing. Then, the study and use of GPU and CUDA processing are included in Section 2.5. The needs of embedding parallelism in image processing are covered in Section 2.6. The Conclusion of the chapter is discussed at the end Section 2.7.

2.2 MEMORY ARCHITECTURE IN PARALLEL COMPUTING

Based on the ability to process a number of instructions and data simultaneously, the computers have been classified as follows [6, 7]:

- Single instruction, single data (SISD);
- Multiple instruction, single data (MISD);
- Single instruction, multiple data (SIMD); and
- Multiple instruction, multiple data (MIMD).

2.2.1 SINGLE INSTRUCTION, SINGLE DATA (SISD)

The name itself suggests that it can execute only one instruction only on single data stream. These basically are uni-processor machines. These machines follow the von Neumann architecture. Figure 2.2 shows the von Neumann architecture [8].

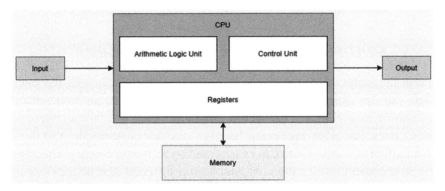

FIGURE 2.2   von Neumann architecture [8].

The von Neumann architecture mainly had four components:

1. **Control Unit:** It generates various types of control signals like op-code fetch, memory read, memory write, I/O read, I/O write. To complete each cycle, the corresponding control signal is generated by the control unit.
2. **Arithmetic Logic Unit (ALU):** It carries out all the arithmetic and logical operation.
3. **Memory Unit:** The data is stored in this unit.
4. **I/O Devices:** The input and output of data is carried out via these devices.

These components process the instructions in a sequential manner. All the programs are executed by following the von Neumann execution cycle, also known as the fetch-decode-execution cycle. This cycle includes the following steps:

1. **Fetch:** The control unit determines the location of the next program instruction to be executed through program counter (program counter, or PC, is a register present in the CPU which stores the address of the next instruction to be executed) and fetches it from the main memory.
2. **Decode:** The instruction is then decoded into the machine language format. And the required data operands for executing the

instruction are fetched from the main memory and stored in the registers of the CPU.

3. **Execute:** Finally, the instructions are executed by the ALU and the results are stored in the registers or in the main memory.

2.2.2   MULTIPLE INSTRUCTION, SINGLE DATA (MISD)

From the name we can derive that the multiple instructions on the same data can run simultaneously. This type of computer has an arrangement as shown in Figure 2.3. In such a computer, there are "n" numbers of processors. For every processor there is a separate control unit. And there is a unified memory which can be accessed by all the processors. This type of arrangement allows the computer to run different instructions simultaneously on the same data. This happens as the processors receive the same data from the memory and different control units can give different instruction to their corresponding processors. This is nothing but instruction level parallelism.

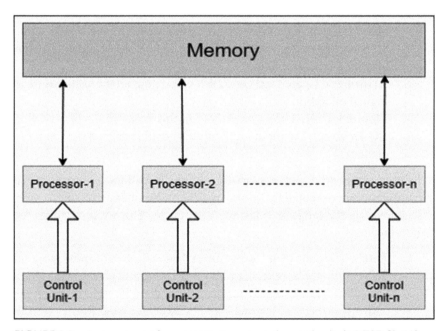

FIGURE 2.3   Arrangement of processors, memory, and control units in MISD [9, 11].

2.2.3 SINGLE INSTRUCTIONS, MULTIPLE DATA (SIMD)

We can easily make a guess of MISD type of computers and its arrangement of components after reading about the SIMD type of computers. Figure 2.4 shows us the arrangement of components in MISD type of computers.

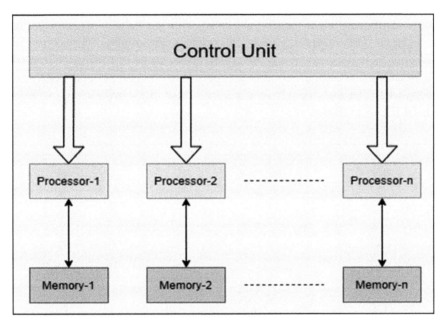

FIGURE 2.4 Arrangement of processors, memory, and control units in SIMD [9, 11].

Computers in this category have a single control unit. Hence all the processors receive the same control signal. Each processor has its own main memory, which means they all can contain data that are different from one another. When the instructions are executed by the processor, the instruction is executed on different data. This corresponds to data level parallelism. These types of computers are widely used. Supercomputers were also built based on this structure.

2.2.4 MULTIPLE INSTRUCTIONS, MULTIPLE DATA (MIMD)

Multiple instructions, multiple data (MIMD) type of computer is perhaps the most interesting and most powerful type of computer among all the four. Here the computers have the same number of processors, data streams and instruction streams. And they are multiple in numbers. Each processor is controlled by its own control unit (see Figure 2.5). So, as stated the data stream and the control unit of all the processors are different, these can work on different data with the instructions also being different. In simple terms it can run different programs in the same time. This parallelism can be achieved with the help of threads and/or processes. This means that they can run asynchronously. These types of computers in the present day are widely used. Some of the most common applications are super computers, personal computers, and computer networks. The only disadvantage here is that asynchronous algorithms are difficult to design.

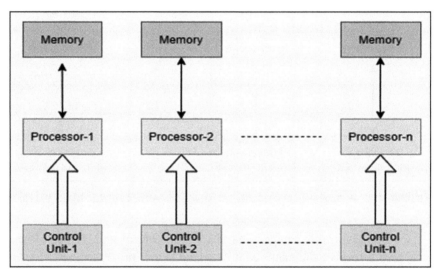

FIGURE 2.5   Arrangement of processors, memory, and control units in MIMD [9, 11].

Now, one problem at this stage that we must dig into is the organization of the memory. Memory organization is important as the data is stored in the memory and the speed at which the memory is accessed must be comparable to the speed of the processor. The speeds must be comparable as the speed of any processor is usually very high. How do we compare the speed of the access of memory and the speed of the processor? We need a quantity of comparison. This comparison is done with the help of the memory cycle time and the cycle time of the processor. Any cycle time can be defined as the time between two successive operations, i.e., memory operations or operations performed by the processor. If these speeds are not comparable then it slows down any process that the computer runs.

The organization of memory [1] has been divided into three types:

- Shared memory;
- Distributed memory; and
- Hybrid memory.

Shared Memory corresponds to a unified memory to which all the processes have an equal access to it. And distributed memory corresponds to the memory which is possessed by every processor individually and other processors do not have an access to it. These organizations of memory differ in the structure of virtual memory or the memory from the perspective of the processor. The distributed memory is often physically distributed. What this means is that physically the memory is unified but internally it is distributed and each part is of the distributed memory is restricted to be accessed by only a certain processor.

The major difference in these organizations of memory is the way they are accessed. Let us assume a memory location 1000. Considering the shared memory organization first, if there are two processors that have an instruction to read from the memory location 1000 at the same time then both the processors read the same data from the same location as the memory is unified and accessible by both the processors. Hence, read from the same data from the same memory. On the other hand, if it were distributed memory then every processor has its own local memory. Now if the same instructions are performed by the processors, i.e., to read from the memory location 1000, then the processor perform this operation using their local memory. Both the processors in this case might read different data from the memory locations as the memory from which they are reading is different.

2.2.4.1 SHARED MEMORY

Shared memory architecture [11] is depicted in Figure 2.6. The system bus allows the transfer of data between the processors and the shared memory. There is a cache memory associated to every processor. These cache memories work as any usual cache memory. The cache memory stores the data that is in the local memory which has a higher probability to be accessed or used by the processor. Now, if any of the processors have to modify some data at a particular location of the shared memory then what happens? The processor first makes the necessary changes at the memory location and then sends a message to all the other processors to notify them about the modified data. This is regarded as the problem of cache coherence. This is nothing but a special case of memory consistency.

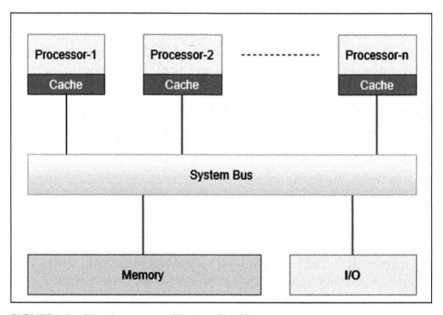

FIGURE 2.6   Shared memory architecture [1, 11].

Synchronization in such architecture becomes a challenging task. It is so because if the synchronization is not done properly then the problem of memory consistency cannot be handled. The accessing of the shared memory is controlled so as to synchronize the processors correctly. Hence only one processor can access the shared memory at a time. Also, the shared memory location must not be changed if another processor or another task is busy accessing it. Different methods are used to access the memory by different computers as per the use cases. Some of the methods are uniform memory access (UMA), non-uniform memory access (NUMA), no remote memory access (NORMA), cache only memory access (COMA).

2.2.4.2   DISTRIBUTED MEMORY

Each processor, is this case, is associated with memory and I/O devices. The memory here is usually logically distributed, but often it means that it is physically distributed too. The distributed memory architecture is depicted in Figure 2.7. If we consider a single unit, i.e., a processor and the cache, memory, and I/O devices associated to it, and then this can make up a separate computer on its own. This is why this kind of architecture is also called 'multicomputer.'

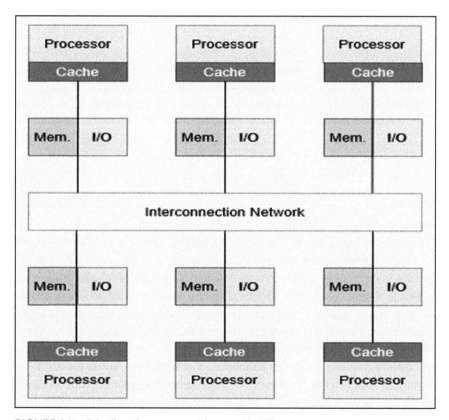

FIGURE 2.7   Distributed memory architecture [1, 11].

The distributed memory architecture [1] has a few pros in comparison with the shared memory architecture. The shared memory architecture has a bus for the communication and data transfer between the processor and the memory. Here each processor has its own memory so the problem of sharing the bus does not exist as the processor communicated directly. Common bus imposes an intrinsic limit to the number of processors that can be connected to it which is not valid in this case as the bus does not exist here. The problem of cache coherency is reduced here as the memory associated with each processor is separate and they work with it only. One problem that does occur in this architecture is the problem when two processors have to communicate between them. This uses the interconnection network for the communication and the communication slows down the task that they are performing. This happens because once the processor starts communication the focus of the processor is shifted to building and sending the message. Hence slowing down both the processors that are communicating with each other. A message passing protocol (MPP) is followed by the processors to communicate or exchange data between them. The communication happens between them through the exchange of data packets. Synchronization is this case is achieved by movement of data/messages between the processors.

2.2.4.3   HYBRID MEMORY

A hybrid architecture [1] is a mixture of both the shared memory and distributed memory. In this a number of systems with share memory architecture are connected via an interconnection network live a distributed memory. This is shown in Figure 2.8. In the present day this kind of architecture is fairly easy to understand because the computers being built have multiple sockets with each socket allowing a multi-core CPU in it is built up as a share memory machine. There are varied numbers of paradigms used for hybrid parallel programming in the software's viewpoint. Open MP with MPI is one such paradigm. MPI is used to communicate among the different shared program machines and the Open MP is used for the communication of the processors in a single unit of shared computer itself. One modification can still be done to this kind of architecture. The CPUs in a single unit of shared memory machine can be replaced by GPUs. Such a modification increases the computing capability of the machine. With this small change there comes a change even in the paradigm of programming. In such case there is use of CUDA/OpenCL along with OpenMP and MPI. CUDA/OpenCL is mainly for the GPUs. We discuss about GPU and CUDA in a section further in this chapter.

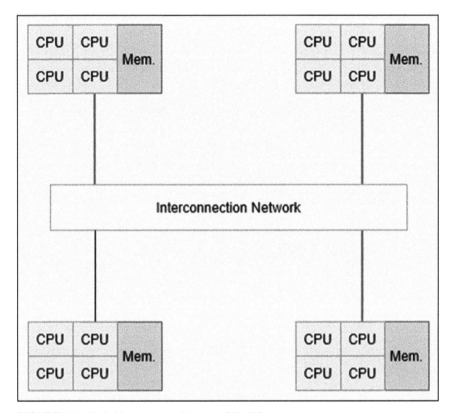

FIGURE 2.8   Hybrid memory architecture [10, 11].

2.3   TERMINOLOGY IN PARALLEL COMPUTING

2.3.1   FLYNN'S TAXONOMY

Flynn in the year 1966 published a seminal paper in the Proceeding of IEEE about the taxonomy of computer architectures. He divided them into 4 categories based on the number of data and instructions that can be simultaneously processed by a computer. This categorization of computers has been discussed in details previously-SISD, MISD, SIMD, and MIMD in Section 2.2. A representation of these types of computers in the form of a graph, to easily remember them, with data and instructions as their axes has been shown in Figure 2.9.

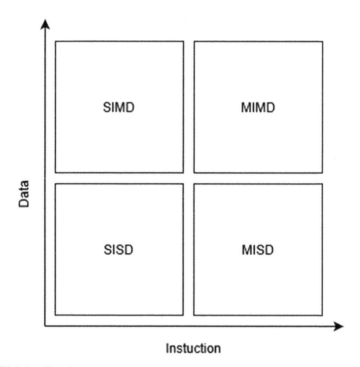

FIGURE 2.9   Flynn's taxonomy.

2.3.2   VON NEUMANN ARCHITECTURE

The Von Neumann architecture was discussed earlier and is depicted in Figure 2.2. The earliest computer such as the ENIAC was designed as a program-controlled machine in which for writing a new program or to change some existing programs rewiring, redesigning, and restructuring was required for which programming was a very tedious task. To ease these programming difficulties the stored program concept was proposed by John Von Neumann. It states that both data and instructions are to be stored in the main memory. Thus, a computer can get instructions from the memory itself and the programs could be changed within the memory itself without redesigning the computers. Von Neumann Architecture was one among the first architectures to be built and the same was also used to try to achieve a certain level of parallelism.

The parallelism was not exactly what we see today in computers with two or more processors. Here the parallelism can be considered to be illusion of parallelism but what actually happens is that there is a quick context switch among the programs that are present in the main memory that are ready for being executed or processed. The processor runs a program and halts based on the algorithm used and takes up another program from the main memory and runs it or resumes running it if it was halted previously. Some of the algorithms designed for uni-processor machines are round robin, first come first serve, shortest job first and longest job first. There are many more such algorithms.

2.3.3   SPEED UP

In order to define speed up we have to define two other quantities-parallel run time and serial run time.

Parallel run time, referred to as *TP*, is defined as the time taken from the beginning of the parallel computation to the end of the computation done by the last processor. The serial run time, referred to as *TS*, is defined as the time taken to execute a program in a serial manner. Now we can formally define the speed up. Speed up is defined as the serial run time of the best sequential algorithm to solve the problem to the parallel algorithm to run the same problem on "p" processors as in Eqn. (1).

$$S = \frac{T_S}{T_P} \tag{1}$$

Based on this definition of speed up, efficiency has also been defined in Eqn. (2). Efficiency (E), is defined as the ratio of speed up (S) to the number of processors (p).

$$E = \frac{S}{p} = \frac{T_S}{p T_P} \tag{2}$$

2.3.4   SCALABILITY

Every computer has a scope to perform better with respect to speed up than how it is performing presently. The ability of a computer to show an increase in its speed up by the inclusion or addition of more resources is called scalability. The factors that give a considerable contribution to this are as follows:

- Hardware such as memory, CPU, and GPU;
- Algorithms used;
- Parallel overhead related;
- Application's features.

2.3.5 PARALLEL OVERHEAD

Parallel overhead is the time that is needed to coordinate the parallel tasks that are ready to run, as opposed to doing useful work. The factors that are included in this are as follows:

- The time needed to start a task;
- Synchronizations;
- Data communication;
- Overhead imposed by software, i.e., by parallel languages, libraries, operating system (OS);
- Time needed to terminate a task.

2.4 DIFFERENT PARALLEL COMPUTING MODELS

Parallel computing models [1, 12] can be of either of two categories: (1) process interaction; or (2) problem decomposition. Process Interaction itself is divided into three: (i) shared memory; (ii) message passing; and (iii) implicit interaction. Problem decomposition is divided into three as well: (a) task parallelism; (b) data parallelism; and (c) implicit parallelism.

A tree on this categorization has been depicted in Figure 2.10.

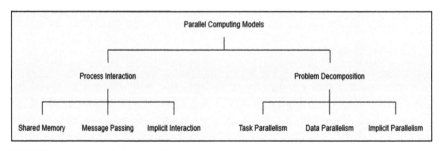

FIGURE 2.10   Parallel computing models classification [1, 11].

2.4.1   PROCESS INTERACTION

This refers to the way in which the interactions among the parallel processes take place. The interactions can be three forms, i.e., shared memory, message passing and implicit interactions. Implicit interactions are those that are not visible to the user/programer.

1. **Shared Memory:** We can relate this to the topic of shared memory machines that was mentioned earlier. Here there is a memory that is shared among the processors which is accessed by them asynchronously. The data and the program are available in this memory. The program is build-up of different processes. Each processor is assigned a different process with the related data to it. The processes are executed or processed by the respective processors and at the end all the processes are re-joined back to the main program.

   Now, this kind of programs can lead to problems like deadlock and race conditions. Deadlock is a situation where the two more processes are unable to progress further as they are in turn waiting for some other processes to finish which in turn are again waiting for some other processes to finish and so on. And race condition is a situation where two or more processes try to access and modify the same data almost at the same time. The final results might not be the desired results in this case. However, multi-core processors support shared memory concept with languages and libraries like OpenMP.

2. **Message Passing:** It refers to the passing of messages from one process to another. There are mainly two different message passing modes: synchronous and asynchronous. Synchronous message passing includes the passing of messages between two processes where the receiver must be ready to receive the message. In an asynchronous mode the receiver is not ready for receiving the message. In here the processes assigned to a single processor might be more than on. The processes that need to send or receive data use certain instruction. For example, we can consider synchronous mode and the instructions used to send and receive data could be send() and receive(), respectively. As it is synchronous mode, every receive() instruction needs a send() instruction. This is illustrated in Figure 2.11.

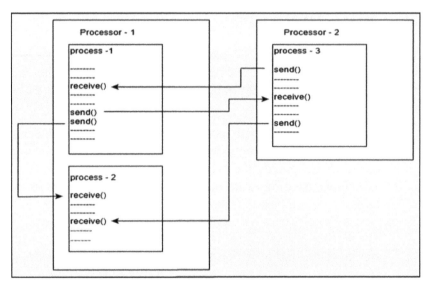

FIGURE 2.11   Message passing [1, 11].

3. **Implicit Interaction:** There are certain situations where the message passing occurs between two processes but the user/ programer is not able to see it. I usually happen when the compiler or runtime are responsible for such passing of messages. They usually occur with two kinds of languages, which are domain specific languages and functional programming languages. Domain specific languages are those which are specialized to a specific domain and implicit interaction occur here because concurrency in high level operation is a common phenomenon. Functional programming languages are those which are constructed only by the composition of functions. The implicit interactions are here present when there is no side-effect of which allows the function's execution be parallel. Managing such parallelism is difficult but there are languages such as Concurrent Haskell which make it possible to manage such parallelism by the user/programer.

2.4.2   *PROBLEM DECOMPOSITION*

Any problem that is solved using programming capabilities can be divided into sub-problems. These sub-problems are not related to each other and can be run separately. Hence these sub-processes can be run in parallel with one another:

1. **Task Parallelism:** Task here refers to a process or thread. A program can be divided into sub-processes or threads. These sub-processes or threads are independent of each other which emphasize the passing of messages. This naturally expresses message passing parallelism. Also as mentioned earlier, MISD, and/or MIMD are allows task parallelism in Flynn's taxonomy. As said above a program can be divided into threads, these threads might be run by the same processor considering the two threads as different processes and using context switch among themselves. Otherwise, the two threads can be run by different processors as well. This has been illustrated in Figure 2.12.

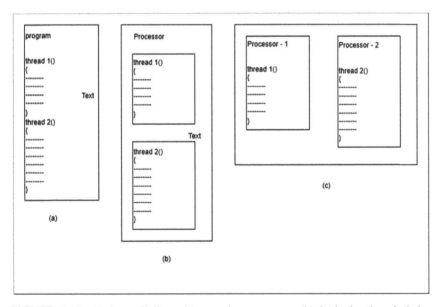

FIGURE 2.12 Task parallelism: (a) complete program; (b) both the threads being processed by a single processor as two different processes by context switching; (c) both the threads processed by two different processors [1, 11].

2. **Data Parallelism:** This was earlier discussed in Flynn's taxonomy under SIMD and/or MIMD. The data parallelism focuses on running a set of operations on some data. The data is usually in the

form of a structured array. The task is independently performed on the data in disjoint partitions. Many programs might want to operate the same operation on different data. This is when data parallelism works the best. For example, we need to multiply a set of 100 numbers to another set of 100 numbers and save all the 100 results then data parallelism could be used very wisely. Both the sets of 100 numbers that need to be multiplied is saved as an array 'a[100]' and 'b[100]' and storing the result in another array named 'r[100].' If a program is written to multiply a set of 25 numbers with another set of 25 numbers, then this could be used in the case of multiplying 100 numbers using data parallelism. We do so by making 4 disjoint sets of 25 numbers each and then passing them on to 4 different processors to execute the multiplication. Finally, all the results are combined. This is illustrated in Figure 2.13.

3. **Implicit Parallelism:** We discusses in the implicit process interaction that such interactions are not visible to the user/programer. The same is true here also. Implicit parallelism is not visible to the user/programer. The compiler and/or the runtime and/or the hardware are responsible for such parallelism. An example that can be given for this is: the compilers make use of automatic parallelization while converting sequential code to parallel code.

2.5 GPU AND CUDA

2.5.1 BASIC OF GPU

Graphical processing Unit (GPU) became an integral part of computing now a day with increased number of image data for computation. In the modern computing age, GPUs are used in every personal computer, laptop, desktop, workstations, mobile phones, game consoles and embedding systems as a multi-core and multi-threaded multiprocessor. The fields of study which related to visual processing like image processing, computer graphics and computer vision are prominently use GPU for processing of its applications. The high processing capacity of GPU is being credited by heavily parallel processing units.

So, the GPUs have ignited a worldwide artificial intelligence (AI) boom. Hence, they have become a key part of modern supercomputing.

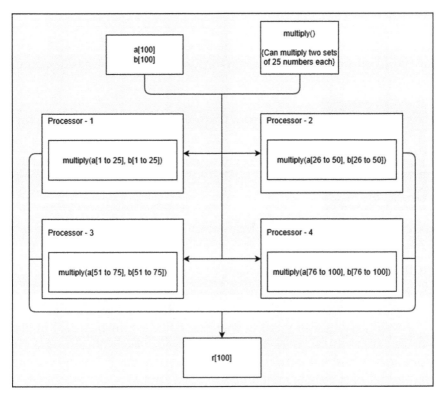

FIGURE 2.13   Data parallelism [1, 11].

2.5.2   CPU VS. GPU

In the initial days of computing, central processing units (CPUs) are largely used because of its popularity. Even if CPU can process the visual data computing, still it has some limitation. While GPUs are having thousands of cores present in it, CPUs have limited. Hence CPUs are good for serial processing, whereas GPUs are preferred for parallel processing. Figure 2.14 shows the core difference between the two processors.

2.5.3   THE GPU ARCHITECTURE

A single GPU consists of multiple processing clusters, each having multiple streaming microprocessors (SMs). Each of these SMs accommodates multiple threads, multiple processor cores, a L-1 cache layer, and a L-2 shared cache. Each processor core inside the SM executes instructions for parallel threads.

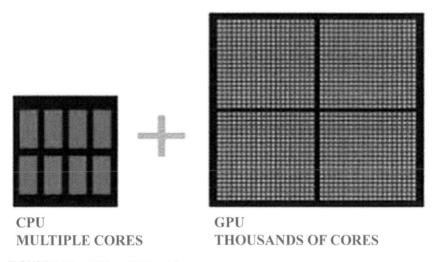

CPU
MULTIPLE CORES

GPU
THOUSANDS OF CORES

FIGURE 2.14   CPU vs. GPU architecture.

Source: Reprinted from Ref. [13]. https://creativecommons.org/licenses/by/3.0/

To address different market segments, GPUs are often scaled by the number of processor cores and memories. This enables them to use the same architecture and software. In case of NVIDIA, this is done by varying the number of streaming multiprocessors and DRAM.

Figure 2.15 shows the schematic architecture of NVIDIA's GPU. It composed of several number of SIMD multi-processor. Each processor included a tiny but fast *shared memory* of approximately 16 to 48 KB. All these processors in the multiprocessor have a direct access to this shared memory. Additionally, all multiprocessors have access to three other device-level memory modules: the *global, texture*, and *constant* memory modules, which are also accessible from the host. The global memory supports read and write operations and it is the largest with size ranging from 256 MB to 4 GB. The texture and constant memories are much smaller and offer only read access to GPU threads. Apart from size, the critical characteristic differentiating the various memory modules is their access latency. While accessing the shared memory takes up to four cycles, it takes 400 to 800 cycles to access global memory. Consequently, to achieve maximum performance, applications should maximize the use of shared memory and processor registers [14].

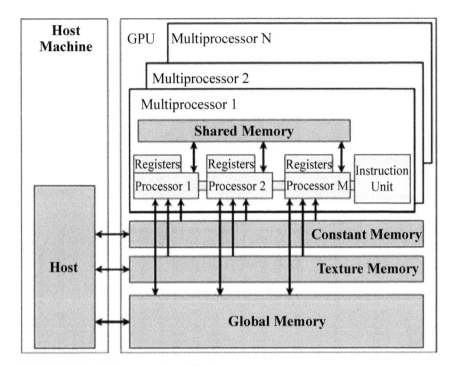

FIGURE 2.15   GPU schematic architecture.

Source: Adapted from Ref. [14].

Now, we will have some highlights in the history and recent development of the GPU technology. Before 1999, most of the graphics processors were non-programmable. The first specialized GPU was NEC µPD7220, implemented as an LSI chip. The term GPU was coined by Sony in reference to the 32-bit Sony GPU in the PlayStation. Table 2.1 illustrates list of NVIDIA's developed products of GPU and their characteristics.

2.5.4   ACHIEVING PARALLELISM: CUDA PROCESSING

The fundamental strength of the GPU is its extremely parallel nature. The CUDA programming model [15] allows developers to exploit that parallelism by writing natural, straightforward C code that will then run-in thousands or millions of parallel invocations, or threads. To utilize the GPU, CUDA provides an extended C-like programming language and compiling. Any call to a global function is said to issue a kernel to run on the GPU and must specify the dimension of the dimension of the grid

TABLE 2.1 NVIDIA's GPU Development

Year	Product/Architecture	Number of Transistors	Number of CUDA Cores	Technology
1997	Riva 128	3 million	—	3D graphics accelerator
1999	GeForce 256	25 million	—	First GPU, programed with DX7 and OpenGL
2001	GeForce 3	60 million	—	First programmable shader GPU, programed with DX8 and OpenGL
2002	GeForce FX	125 million	—	32-bit FP programmable GPU with Cg programs, DX9 and OpenGL
2004	GeForce 6 series	222 million	—	32-bit FP programmable GPU, GPGPU Cg programs, DX9 and OpenGL
2005	GeForce 7 series	302 million	—	64-bit FP programmable GPU, with DX9, GPGPU, and OpenGL. The last generation available on AGP cards
2006	GeForce 8 series	681 million	128	First unified graphics and computing GPU, programed in C using CUDA. Introduced Tesla architecture, which has a unified shading architecture.
2007	GeForce 9 series	754 million	128	Added PCIe 2.0 support, improved color, and z-compression.
2009	GeForce 200 series	1.4 billion	240	Introduced second generation of Tesla microarchitecture. Has double precision support for GP-GPU applications.
2010	GeForce 400 series	3 billion	480	Introduced Fermi microarchitecture and added support for OpenGL 4.0, Direct 3D 11.0 and C++.
2011	GeForce 500 series	3 billion	480	These provided a significant improvement over the 400 series, with all stream processors enabled.
2012	GeForce 600 series	3.5 billion	1,536	Introduced the Kepler microarchitecture, increasing power efficiency. NVEnc was also introduced.
2013	GeForce 700 series	7 billion	2,880	DX12 support was added.

TABLE 2.1 *(Continued)*

Year	Product/Architecture	Number of Transistors	Number of CUDA Cores	Technology
2015	GeForce 900 series	9 billion	2,048	Introduced Maxwell architecture, which has native shared memory atomic operations for 32-bit integers, native shared memory 32-bit and 64-bit compare-and-swap, increased maximum active threads, VXGA, and CUDA compute ability 5.2.
2016	GeForce 10 series	15.3 billion	3,584	Introduced Pascal microarchitecture, which added FP16 support, improved FP64 atomics, HBM2 memory, CUDA Compute Ability 6.0, Unified Memory and NVLink.
2017	Titan series	21.1 billion	5,120	Introduced the Volta microarchitecture, which added NVLink 2.0. Chips also featured 640 Tensor cores, giving higher performance for deep learning applications.
2018	GeForce 20 series	18.6 billion	4,608	Introduced the Turing microarchitecture, with ray tracing support and enhanced tensor core performance with INT4 performance.

and the blocks that will be used to execute the program. Now, Figure 2.16 will depict the difference between the standard C code snippets with its corresponding CUDA version.

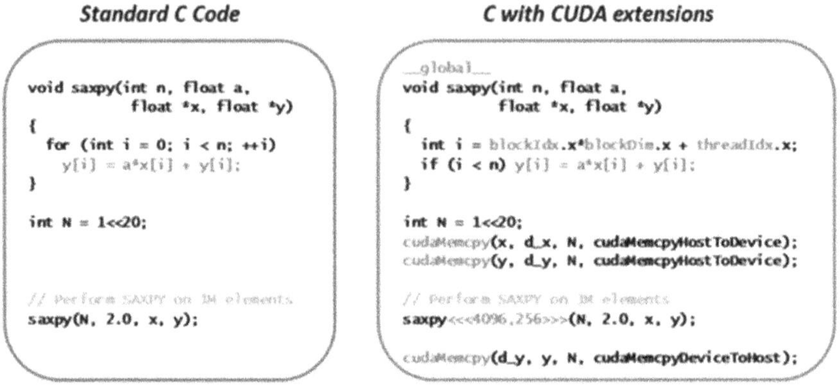

Standard C Code	C with CUDA extensions

```
void saxpy(int n, float a,
           float *x, float *y)
{
  for (int i = 0; i < n; ++i)
    y[i] = a*x[i] + y[i];
}

int N = 1<<20;




// Perform SAXPY on 1M elements
saxpy(N, 2.0, x, y);
```

```
__global__
void saxpy(int n, float a,
           float *x, float *y)
{
  int i = blockIdx.x*blockDim.x + threadIdx.x;
  if (i < n) y[i] = a*x[i] + y[i];
}

int N = 1<<20;
cudaMemcpy(x, d_x, N, cudaMemcpyHostToDevice);
cudaMemcpy(y, d_y, N, cudaMemcpyHostToDevice);

// Perform SAXPY on 1M elements
saxpy<<<4096,256>>>(N, 2.0, x, y);

cudaMemcpy(d_y, y, N, cudaMemcpyDeviceToHost);
```

FIGURE 2.16   Standard C code vs. CUDA extensions.

Source: Adapted from Ref. [15].

2.6   NEED OF EMBEDDING PARALLELISM IN IMAGE PROCESSING TECHNIQUES

Both parallel processing and Image processing being in individual terminology, are two major contributions to the field of computer science. As we have discussed earlier, parallel processing is the study of parallel execution of a task by exploiting the architecture of parallel computing and at the same time, Image processing is the study and processing of images to exploit various feature available in image. The various image processing techniques begins with image acquisition, then various image pre-processing techniques such as image enhancement, noise reduction, binarization, and morphological operation. The important technique includes color image processing, image restoration, image compression, image classification, image localization, image segmentation, image representation, etc. But the inter-disciplinary studies again hold major breakthrough in the field of research and bringing excellent throughput to the study. In this aspect, embedding of parallelism into various image processing techniques emerges as parallel and distributed image processing will provide us state-of-the-art results to this new domain of inter-disciplinary study [15, 16].

In recent past, with the huge increase in the multimedia data in various internet-based applications, other multimedia-based applications and even if in social media websites like Facebook, Twitter, YouTube, e-commerce websites, etc. For example, in the e-commerce website, a particular product needs varieties of images from different views for proper recognition of that product and like that imagine a lots products need more and more number of images and there are again various e-commerce websites. Similarly, there are millions of users on social media and there are also number of social media are available now-a-days. Response to such complex multimedia data can be processed immediately by parallel or distributed image processing. As images contain high dimensional data, normal processor cannot handle these huge amounts of data. So, the introduction of parallel processor to handle these huge amounts of multimedia data comes into the scenario.

Again, images captured may be low intensity images due to surveillance system on road, house, industries; organization is really challenging and complex. So high speed image processing and real time image processing are highly required in this scenario. In real-time image processing, processing should be done with high speed on rapid sequence of images or on single image. Such high-speed image processing can be handled using parallel or distributed image processing. Apart from the above-mentioned application of image processing, it also highly recommended in engineering, medical, space, and satellite, agriculture, and manufacturing industry. These fields of image processing need heavy computation.

Looking to the architecture of the current standalone computer which has single CPU structure, has it won limitation in processor speed, memory capacity and other hardware components. To process the high-end image application, more number of processor and multi-core processor are required, which can execute the application in parallel fashion.

To achieve the result in the area of parallel and distributed image processing field, the various tools, technologies, applications are used. In particulars, the various important tools employed to achieve parallelism in image processing is GPU, CUDA (computed unified device architecture), MPI (message passing interface), Java, Hadoop, and OpenCV and MATLAB. The following Table 2.2 summarizes the tools and technologies used for implementing parallel and distributed image processing. Along with parallel processing of image data, with the recent trending of AI based neural network, machine learning (ML) and deep learning (DL) technologies will be greatly helpful in processing the high-end multimedia data.

2.6.1 APPLICATION OF EMBEDDING PARALLELISM IN IMAGE PROCESSING TECHNIQUES

1. **Medical Image Analysis:** It is always a denting task to handle medical images as we have to maintain the privacy of patients, while handling these data. The various medical imaging modalities include X-rays, MRI, CT, PET, etc. All these modalities of medical image contain patients' information may be in encoded format to hide the privacy of patient on first-hand information. As these medical image data are in encoded format and also are the 3D images, it consumes more memory hence need more complex algorithms to handle. To run these complex algorithms, we also need HPC machine. All the high-performance computing machines are formed with high end hardwires, i.e., GPU with CUDA architecture. So, to handle these heavy volumes of medical image we need GPUs with huge amount of system memory. Sometimes the pre-processing of the medical images also needs complex algorithms which in turns need heavy computation [17–19].

2. **Surveillance System:** This mainly includes the visual tracking images captured by the camera. It is basically is used for security purpose and mostly installed on the industry, the organization, the highways and roads, shopping malls, etc. A huge number of visuals are generated by these surveillance systems as frames of images from video and may the visuals be not in the perfect condition due to its situation and position of capturing. That is the visuals may be having noise, dark spots due low intensity, blur, etc. To process these millions of visuals on a real time basis to maintain the accuracy of security, high computing machine with parallel processor are the compulsory requirement. The processor has to so efficient it should detect the real objects from the whole image by considering the texture, edge, color, gradient, spatial-temporal features of the images. The execution must follow the real time and distributed system to give correct results.

3. **Image Mining:** In this modern era of computation and internet, there is a rapid increase in multimedia data specifically the image data in various internet-based application and standalone application such as various e-commerce sites, social media sites and search engines. To find out the pattern from these huge amounts of image data by the image mining to enhance the business strategies

TABLE 2.2 Tools and Technologies Used in Parallel and Distributed Image Processing Implementation [16]

Tool and Technology	Brief Detail of Tool and Technology	Benefits of Implementation
GPU with CUDA programming	Computing device with NVIDIA GPU	Used to achieve parallelism and speedup and good throughput.
MPI (Message passing interface)	Specification of communication protocol. It acts as a language and platform independent.	Execution time is drastically changed by utilizing image correlation quality.
OpenMP	De-facto standard API for shared memory parallel applications	Helpful in multicore architecture, achieve parallelism.
OpenCV	Highly optimized library with focus on real-time applications	Motion, edge, line detection
Java	Java programming language	Multithreading using Java on multi core computer. Good speed up and parallel efficiency with suitable multithreading.
Clik	ANSI C based general-purpose programming language	Multithreaded parallel computing.
MATLAB with parallel computing toolbox	Parallel environment for MATLAB programs.	Parallel processing on multi-core processors and GPU.

is always a big step towards positivity. But the big challenge is always lying with to handle and process this heavy amount of image data. The parallel environment for image data mining (PEIDM) is an environment that was designed to exploit the correct pattern from these huge amounts of data.

2.7   CONCLUSION

In this chapter, we make exploratory study of parallel computing with architecture and application in depth. We also discuss all the necessary tools and terminologies related to parallel processing for further enhancing the knowledge of reader and developer. The overall study suggests that parallel computing helps to save the time of computation and solve larger problems. So parallel computing is fast and future of computing. We can incorporate the all the current trends and technologies in the parallel computing study make it the most inter-disciplinary study to provide the state-of-the-art performance.

KEYWORDS

- cache only memory access
- graphical processing unit
- message passing protocol
- non-uniform memory access
- remote memory access
- uniform memory access

REFERENCES

1. Barney, Blaise. "Introduction to parallel computing." *Lawrence Livermore National Laboratory* 6.13 (2010): 10.
2. Kumar, Vipin, et al. *Introduction to parallel computing*. Vol. 110. Redwood City, CA: Benjamin/Cummings, 1994.
3. Golub, Gene H., and James M. Ortega. *Scientific computing: an introduction with parallel computing*. Elsevier, 2014.
4. Dongarra, Jack, et al. *Sourcebook of parallel computing*. Vol. 3003. San Francisco^ eCA CA: Morgan Kaufmann Publishers, 2003.
5. Hwang, Kai, and A. Faye. "Computer architecture and parallel processing." (1984).

6. Carter, John B., et al. "Avalanche: A communication and memory architecture for scalable parallel computing." *Proc. of the Fifth Workshop on Scalable Shared Memory Multiprocessors*. 1995.

7. Vishkin, Uzi. "Computer memory architecture for hybrid serial and parallel computing systems." U.S. Patent No. 7,707,388. 27 Apr. 2010.

8. Hur, Rotem Ben, and Shahar Kvatinsky. "Memory processing unit for in-memory processing." *2016 IEEE/ACM International Symposium on Nanoscale Architectures (NANOARCH)*. IEEE, 2016.

9. Gutzmann, Michael M., and Ralph Weper. *Classification approaches for parallel architectures*. Inst. für Informatik, Lehrstuhl für Rechnerarchitektur und-kommunikation, 1996.

10. Göhringer, Diana, et al. "A taxonomy of reconfigurable single-/multiprocessor systems-on-chip." *International Journal of Reconfigurable Computing* 2009 (2009).

11. Zaccone, Giancarlo. *Python parallel programming cookbook*. Packt Publishing Ltd, 2015.

12. Rajasekaran, Sanguthevar, and John Reif, eds. *Handbook of parallel computing: models, algorithms and applications*. CRC Press, 2007.

13. Saxena, Sanjay, Shiru Sharma, and Neeraj Sharma. "Parallel image processing techniques, benefits and limitations." *Research Journal of Applied Sciences, Engineering and Technology* 12.2 (2016): 223-238.

14. Al-Kiswany, Samer, Abdullah Gharaibeh, and Matei Ripeanu. "GPUs as storage system accelerators." *IEEE Transactions on parallel and distributed systems* 24.8 (2012): 1556-1566.

15. Nickolls, John, and William J. Dally. "The GPU computing era." *IEEE micro* 30.2 (2010): 56-69.

16. Prajapati, Harshad B., and Sanjay K. Vij. "Analytical study of parallel and distributed image processing." *2011 International Conference on Image Information Processing*. IEEE, 2011.

17. Dongarra, Jack J., et al. "A message passing standard for MPP and workstations." *Communications of the ACM* 39.7 (1996): 84-90.

18. Flynn, Michael J., and Albert Podvin. "Shared resource multiprocessing." *Computer* 5.2 (1972): 20-28.

19. Adve, Sarita V., and Kourosh Gharachorloo. "Shared memory consistency models: A tutorial." *computer* 29.12 (1996): 66-76.

20. Gharachorloo, Kourosh, et al. "Memory consistency and event ordering in scalable shared-memory multi-processors." *ACM SIGARCH Computer Architecture News* 18.2SI (1990): 15-26.

21. Chandra, Rohit, et al. *Parallel programming in OpenMP*. Morgan kaufmann, 2001.

22. Dagum, Leonardo, and Ramesh Menon. "OpenMP: an industry standard API for shared-memory programming." *IEEE computational science and engineering* 5.1 (1998): 46-55.

23. Gropp, William, et al. *Using MPI: portable parallel programming with the message-passing interface*. Vol. 1. MIT press, 1999.

24. Athas, William C., and Charles L. Seitz. "Multicomputers: Message-passing concurrent computers." *Computer* 21.8 (1988): 9-24.

25. Riener, Andreas. *Sensor-actuator supported implicit interaction in driver assistance systems*. Wiesbaden: Vieweg+ Teubner, 2010.

26. Wilson, Andrew, and Nuria Oliver. "Multimodal sensing for explicit and implicit interaction." *11th International Conference on Human-Computer Interaction (HCI International 2005), Las Vegas, Nevada, USA*. 2005.

27. Foster, Ian. "Task parallelism and high-performance languages." *IEEE Concurrency* 3 (1994): 27-36.

28. Subhlok, Jaspal, et al. "Exploiting task and data parallelism on a multicomputer." *Proceedings of the fourth ACM SIGPLAN symposium on Principles and practice of parallel programming*. 1993.

29. Peyton Jones, Simon, et al. "Harnessing the multi-cores: Nested data parallelism in Haskell." *IARCS Annual Conference on Foundations of Software Technology and Theoretical Computer Science*. Schloss Dagstuhl-Leibniz-Zentrum für Informatik, 2008.

30. Bik, Aart JC, and Dennis B. Gannon. "Automatically exploiting implicit parallelism in Java." *Concurrency: Practice and Experience* 9.6 (1997): 579-619.

31. Alexandrov, Alexander, et al. "Implicit parallelism through deep language embedding." *Proceedings of the 2015 ACM SIGMOD International Conference on Management of Data*. 2015.

32. Keckler, Stephen W., et al. "GPUs and the future of parallel computing." *IEEE micro* 31.5 (2011): 7-17.

33. Kirtzic, J. Steven, Ovidiu Daescu, and T. X. Richardson. "A parallel algorithm development model for the GPU architecture." *Proc. of Int'l Conf. on Parallel and Distributed Processing Techniques and Applications*. 2012.

34. Garland, Michael, et al. "Parallel computing experiences with CUDA." *IEEE micro* 28.4 (2008): 13-27.

35. Kirk, David. "NVIDIA CUDA software and GPU parallel computing architecture." *ISMM*. Vol. 7. 2007.

36. Gonzalez, Rafael C., Richard Eugene Woods, and Steven L. Eddins. *Digital image processing using MATLAB*. Pearson Education India, 2004.

37. Jena, Biswajit, Gopal Krishna Nayak, and Sanjay Saxena. "Maximum Payload for Digital Image Steganography Obtained by Mixed Edge Detection Mechanism." *2019 International Conference on Information Technology (ICIT)*. IEEE, 2019.

38. Duncan, James S., and Nicholas Ayache. "Medical image analysis: Progress over two decades and the challenges ahead." *IEEE transactions on pattern analysis and machine intelligence* 22.1 (2000): 85-106.

39. Dhawan, Atam P. *Medical image analysis*. Vol. 31. John Wiley & Sons, 2011.

40. Swanson, Daniel R., Jerry M. Moen, and Bradley M. Tate. "Event surveillance system." U.S. Patent No. 5,689,442. 18 Nov. 1997.

41. Conklin, David Allen, and John Reed Harrison. "Network surveillance system." U.S. Patent No. 5,991,881. 23 Nov. 1999.

42. Mazurkiewicz, Adam, and Henryk Krawczyk. "A parallel environment for image data mining." *Proceedings. International Conference on Parallel Computing in Electrical Engineering*. IEEE, 2002.

43. Zhang, Ji, Wynne Hsu, and Mong Li Lee. "Image mining: Issues, frameworks and techniques." *Proceedings of the 2nd ACM SIGKDD International Workshop on Multimedia Data Mining (MDM/KDD'01)*. University of Alberta, 2001.

CHAPTER 3

BASIC UNDERSTANDING OF MEDICAL IMAGE PROCESSING

PRADEEP KUMAR,[1] SUBODH SRIVASTAVA,[1] and Y. PADMA SAI[2]

[1]*Department of Electronics and Communication Engineering, National Institute of Technology Patna, Bihar, India*

[2]*Department of Electronics and Communication Engineering, VNR VJIET Hyderabad, Telangana, India*

ABSTRACT

In 1895 after the discovery of x-ray, images were commonly used for medical diagnostics. The process of taking images of the body parts for medical uses in order to study or identify various diseases is known as medical imaging. Throw out the world every week there are millions of imaging procedures been done. In this rapidly growing medical imaging process, including image recognition, analysis, and enhancement in image processing techniques. Image processing increases the percentage and amount of detected tissues. In this chapter, the authors presented a basic understanding of medical images and processing. Image processing is often viewed as arbitrarily manipulating an image for achieving an esthetic standard or supporting a preferred reality. Why we do the image process is because the fact is that the human visual system does not recognize the world in a similar way as sensors that record in the form of digital data.

3.1 INTRODUCTION

Medical imaging is the process of producing visible images of inner structures of the body for scientific and medicinal study and treatment as well as a visible view of the function of interior tissues. This process pursues the disorder identification and management. This process creates data bank of regular structure and function of the organs to make it easy to recognize the anomalies [1].

By the increasing use of direct digital imaging systems for medical diagnostics, in health care digital image processing (DIP) is becoming very important. In addition to originally digital methods, such as magnetic resonance imaging (MRI) or computed tomography (CT), endoscopy or radiography comes under analog imaging modalities, this have been rigged with digital sensors. Hence, medical image processing is becoming an essential part of medical technology [2, 3].

Image processing is an approach in which an operation is done on image to get enhanced image or to bring out some useful details from it. It is a kind of signal processing in which contest like input is image and output are also image or it can also be characteristics features of image. Nowadays, Image processing is amid drastically improving technologies. It forms core research platform within engineering and computer science disciplines too. Image processing mainly consists of the following procedure (Figure 3.1):

- Firstly, the image on which we want to do processing has to be imported through tools used for image acquisition;
- Secondly, we have to analyze and manipulate the image based on our requirements; and
- Lastly after the manipulations and the changes done to the image, we have our output image which is the processed one.

Digital images are composed of lone pixels (this acronym is formed from the element and words picture), where having different brightness or color values are assigned. They can be objectively evaluated, accurately processed, and made available at many places at the same time by means of appropriate communication networks and protocols, such as picture archiving and communication systems (PACS) and the digital imaging and communications in medicine (DICOM) protocol, respectively [3, 6]. Based on digital imaging techniques, the entire spectrum of DIP is now applicable in medicine. The commonly used term medical image processing means the provision of DIP for medicine, medical image processing covers four major areas shown in Figure 3.1 [7].

1. **Image Formation:** It is a process which captures an image to forma digital image matrix.

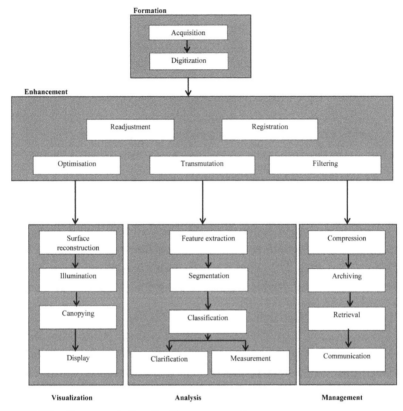

FIGURE 3.1 Component of medical image processing.

2. **Image Visualization:** It refers to all types of manipulation of this matrix, resulting in an optimized output of the image.

3. **Image Analysis:** It includes all the steps of processing, which are used for quantitative measurements as well as abstract interpretations of medical images.

   These steps require a-priori knowledge on the context and content of the images, which must be integrated into the algorithms on a high level of abstraction. Thus, the process of image analysis is very specific, and developed algorithms can be transferred rarely directly into other domains of applications.

4. **Image Management:** It sums all techniques that provide the efficient storage, communication, transmission, archiving, and access (retrieval) of image, since an uncompressed radiograph may require

several megabytes of storage capacity. The methods of telemedicine are also a part of the image management.

5. **Image Enhancement:** In contrast to image analysis, which is also referred to as high-level image processing, image enhancement or low-level image processing denotes automatic or manual techniques, which does not require any pre knowledge to realize on the specific content of images. This type of algorithm has similar effects regardless of what is shown in an image [6].

3.2 CLASSIFICATION OF DIGITAL IMAGES

Namely there are 2 types of Image Processing, Analog, and DIP. Analog image processing can be found useful for the hard copies like printouts and photographs. DIP techniques helps in manipulation of the digital images by using computers. The three general phases that all types of data must undergo while using digital technique are pre-processing, enhancement, and display, information extraction [1–3].

Pixel is a smallest unit of an image. So, generally a digital image is considered as a 2-dimensional array of pixels. There is an intensity possessed by each pixel which is called gray level or gray value. The intensity/gray values are usually denoted in 8-bit integer. So, the range of values lies from 0 to 255. The values which are nearer to 0 represent dark region and the values which are nearer 255 represent bright region.

Based on this gray level, we have three types of images in general, and they are:

1. **Binary Image:** It has 0 and 255 as its feasible gray intensities, there are no mid values. Binary images are used as false face for specifying the pixel of interest in various processing tasks of images.

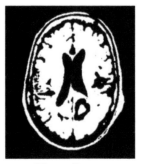

Binary Image

2. **Gray Scale Image:** It has 0 to 255 as range values that are each pixel location can attain any value from 0 to 255.

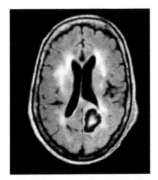

Grayscale Image

3. **Color Image:** Both gray scale images and binary images are 2-D arrays, where at each and every location, only a single value is required to represent a pixel. To create a color image, there is a compulsion for greater than one value for each pixel. There is a requirement of three values for each pixel to represent any color. This 3-value combination to create a color space is called RGB color space.

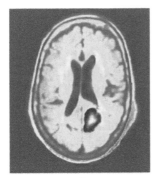

Color Image

3.3 MEDICAL IMAGING SYSTEMS

Medical imaging is the methodology through which visual representation of the interior of a body for clinical analysis and medical intervention is created, as well as visual representation of the function of some organs or tissues is generated.

Some of the common medical imaging system involves [1–4, 6]:

1. **Mammography:** It is a process in which we use low-energy x-rays to examine the human breast, the mammogram is used to detect the breast cancer in early stages in woman.

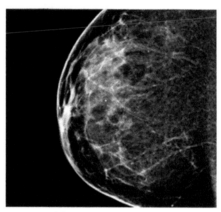

Mammographic Image

An x-ray is a noninvasive medical test that helps physicians to diagnose and treat medical conditions. Imaging with x-rays includes uncovering of a body part to a miner dose of ionizing radiation to generate images of inside of the body. X-rays are the oldest and most frequently used form of medical imaging.

2. **Computed Tomography (CT):** The term "computed tomography (CT)" refers to computer-based x-ray imaging process in which a small beam of x-rays is pointed at a patient and is quickly rotated around the body, which results in generation of cross-sectional images or slices of the body. These slices are known as tomographic image and contain high detailed information than conventional x-rays. After the collection of successive slices by the device, they can be digitally "stacked" together to form a 3D image of the patient that allows for easier identification and location of basic structures as well as possible tumors or abnormalities.

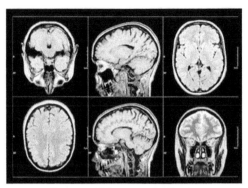

Computed Tomography Image

3. **Ultrasound:** It is a technology, which uses sound waves with frequency higher then upper audible limits of human which returns the echoes from body, these echoes are used to generate image of internal structure. Ultrasound and location determined technique is similar, some animals like whales and bats use this technique. Using a transducer ultrasound will transmit high-frequency pulses inside the body, as this wave travels through the body tissues some waves are absorbed, and some waves are reflected. This reflected wave is received through a transducer and converted into electric signals. These received electric signals are again converted into digital ones and passed through the computer system. The computer system uses the arithmetic and logic calculation to form the two-dimensional images of the scanned structures. In the ultrasonic system, thousands of pulses are sent per each millisecond.

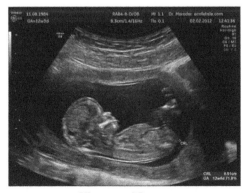

Ultrasound Image

4. **Magnetic Resonance Imaging (MRI):** It is a non-invasive imaging technology which generates 3D detailed anatomical images. It is often used for disease detection, diagnosis, and treatment monitoring. This is based on advanced technology which detects and excites the change in the direction of the rotational axis of protons found in the water that makes up living tissues.

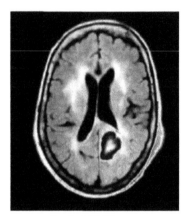

Brain MRI Image

5. **Nuclear Medicine:** It is a medical specialty that uses radioactive tracers (radiopharmaceuticals) to assess bodily functions and to diagnose and treat disease. Specially designed cameras allow doctors to track the path of these radioactive tracers. Single photon emission computed tomography or SPECT and positron emission tomography or PET scans are the two most common imaging modalities in nuclear medicine.

3.4 BASICS OF DIGITAL IMAGE PROCESSING (DIP)

The image processing methodology involves the following steps:

1. **Acquisition of Image and Storage:** This is the fundamental step involved in Image Processing Methodology. This stage of image

acquisition stage involves retrieving an image from various sources like sensors, Open-source repositories are the ones which are made inputs from a scanner, digital cameras or by the help of serial cameras. This image should be of high-quality and with higher resolution, which supports for better image analysis. After acquisition of the image, it is followed by storage of the image in the local drive or any device [9].

2. **Preprocessing of the Image:** After acquiring of the image and storage process then the next step is regarding the Preprocessing of the image. The preprocessing involves the following operations:

    i.   Run a few transformations on the image (sampling and manipulation; for example, grayscale conversion);
    ii. Enhance the quality of the image (filtering; for example, DE blurring); and
    iii. Restore the image from noise degradation.

3. **Segmentation:** The image needs to be segmented in order to extract the objects of interest. Image segmentation is defined as the multiple segmentation of images. In the form of these segments of an image that is more essential and easier to analyze. Segmentation is practiced by scanning the image, pixel by pixel and after a label is given for each pixel, based on if the gray level is greater or lesser than the threshold value.

4. **Output Image:** After using different methods for processing images linked with morphological operation on digital image, the object of interest from the given image is attained.

3.5 CONCLUSION

Images are the method of expression of the data in pictographic form. Pixels are the small elements which form an image. Every pixel has its own value and position. Geometric image signifies an image arithmetically with geometrical primitives such as lines. Image processing techniques is a group of approaches that are used for handling the images by computer. The main objective of segmentation is to divide the image into important portions. Image segmentation works on three separate methods, they are region, border, and edge. Thresholding segmentation uses the threshold value and histogram of pixels. Image edge techniques are used to analyze the images at borders or discontinuing.

The future of DIP involving new intelligence, digital automated robots created entirely by research scientists in various nations of the world. It includes advancements in various DIP applications. Due to the break-through in image processing and other related technologies, there is a high possibility of how new things will be created in the coming new genera-tions, changing the perspective of the way people look towards the tech-nology in the world. Advance researches in the field of image processing and artificial intelligence (AI) will include voice commands, anticipating. The data requirements of governments, converting languages, identifying, and tracking people and things, diagnosing medical conditions, performing operations and surgery, reprogramming defects in human DNA, and autonomous driving including in all formats of transportation. With the development in the power and advancement of modern computing, the synonym of computation has been extended beyond the present limits. In future, image processing technology will be more progressive, and the visual system of man can be imitated.

KEYWORDS

- computed tomography
- digital imaging
- image processing technique
- magnetic resonance imaging
- medical imaging
- medical imaging modalities

REFERENCES

1. Abdallah, Y., (2015). Improvement of sonographic appearance using HATTOP methods. *International Journal of Science and Research, 4*(2), 2425–2430. doi: http://dx.doi.org/10.14738/jbemi.55.5283.

2. Abdallah, Y., (2016). Increasing of edges recognition in cardiac scintigraphy for ischemic patients. *Journal of Biomedical Engineering and Medical Imaging, 2*(6), 40–48. doi: http://dx.doi.org/ 10.14738/jbemi.26.1697.

3. Abdallah, Y., (2011). *Application of Analysis Approach in Noise Estimation, Using Image Processing Program* (pp. 123–125). Germany: Lambert Publishing Press GmbH & Co. KG.

4. Abdallah, Y., & Yousef, R., (2015). Augmentation of x-rays images using pixel intensity values adjustments. *International Journal of Science and Research, 4*(2), 2425–2430.

5. Abdallah, Y., (2011). *Increasing of Edges Recognition in Cardiac Scintigraphy for Ischemic Patients* (pp. 123–125). Germany: Lambert Publishing Press GmbH & Co. SKG.

6. Jain, K., (1989). *A Handbook of Fundamentals of Digital Image Processing*. Prentice Hall of India.

7. Deserno, T. M., (2011). Fundamentals of medical image processing. In: Kramme, R., Hoffmann, K. P., & Pozos, R. S., (eds.), *Springer Handbook of Medical Technology*. Springer Handbooks. Springer, Berlin, Heidelberg.

8. Gonzalez, R. C., & Woods, R. E., (2009). *Digital Image Processing* (3rd edn.). Prentice Hall of India.

9. Madhuri, A. J., (2010). *Digital Image Processing - An Algorithmic Approach*. Prentice Hall of India.

10. Aisha, A., & Ibrahim, M. H., (2010). *Vehicle Detection Using Morphological Image Processing Technique.* IEEE.

11. Prewitt, J. M. S., (1970). Object enhancement and extraction. *Picture Processing and Psychopictorics*. Academic Press, New York.

12. Dewangan, S. K., (2012). *Devnagari Handwritten Signature Recognition Using Neural Network.* Lambert Academic Publications (LAP), ISBN: 978-3-659-26595-2. Germany.

13. Soumen, B., & Gaurav, H., (2011). Topographic feature extraction for Bengali and Hindi character images. *International Journal of Signal & Image Processing, 2*(2), 181–196.

14. Patel, N., & Dewangan, S. K., (2015). An overview of face recognition schemes. *International Conference of Advance Research and Innovation (ICARI-2015).* Institution of Engineers (India), Delhi State Center, Engineers Bhawan, New Delhi, India.

15. Yang, Y., & Wei, L. N., (2007). Microcalcification classification assisted by content based image retrieval for breast cancer diagnosis. *IEEE International Conference on Image Processing, 5*, 1–4.

16. Shailendra, K. D., (2015). Human authentication using biometric recognition. *International Journal of Computer Science & Engineering Technology* (Vol. 6, No. 4, pp. 240–245). ISSN: 2229-3345.

17. Dewangan, S., Gupta, P., Sahu, U. K., & Verma, I., (2012). Realtime recognition of handwritten words using hidden Markov model. *International Journal of Technological Synthesis and Analysis (IJTSA)* (Vol. 1, No. 1, pp. 7–9). ISSN: 2320-2386.

CHAPTER 4

MULTICORE ARCHITECTURES AND THEIR APPLICATIONS IN IMAGE PROCESSING

T. VENKATA SRIDHAR[1] and G. CHENCHU KRISHNAIAH[2]

[1]Department of Electronics and Telecommunication, IIIT Bhubaneswar, Odisha, India

[2]Department of ECE, GKCE, Sullurupeta, Andhra Pradesh, India

ABSTRACT

Multi-core architectures are the widely popular design techniques that are used in most electronic devices nowadays. Having many cores give a quicker and timely accurate response in many applications, especially like medical image processing, which is a more critical and essential area. Integrating many cores into a single chip is a big task as both software, hardware, and firmware must coordinate simultaneously for better yield. Here in this chapter, the authors tried to cover the basics concepts to moderately high levels of abstraction to understand the three coordination (Hardware vs. Software vs. Firmware). To produce accurate and good results, it is required to understand the compatibilities and issues in combining the hardware and software co-design. Because the field of medical image processing critically requires the rate of false-positive as minimum as possible as it deals with lives.

4.1 INTRODUCTION

When people hear the term High-Performance Computing, many of them have some superstitious imaginations like it is a miracle. The ground reality is, it got that strength through a collaboration of small individual processing units, in a hardware point of view, and collective group executions from a software point of view. The man got inspired by nature a lot all the time, most of the innovations were made real with the help of nature. One such an incident in the understanding of High-Performance computer hardware architecture is "ants carrying a big candy or a grain." If one ant cannot handle the whole candy then few ants join collectively (sharing in multi-cores) and fulfill the task to finish in time. If an interrupt occurs immediately, they change or take the new support and path (dynamic reconfiguration). If you understand the theme then, finding a good architecture for High-Performance computing is simple (Figure 4.1).

FIGURE 4.1   Ants collectively moving a big object.

Source: Used with permission. © E. Fonio, O. Feinerman/Weizmann Institut

4.2   SINGLE AND MULTI-CORE ARCHITECTURES

In this context, we get the knowledge of basic processors and its evolution since inception. Intel is chosen here for convenience and other competitors to intel were moderately used, like AMD, IBM, etc. Figure 4.2 shows the single vs. multi-core processors general block diagrams and Figure 4.3 shows the same for IBM.

In the case of a single-core processor, the whole task is carried by the processor alone and each instruction execution is serial. All the exceptions

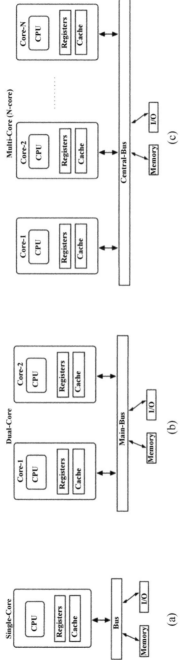

FIGURE 4.2 Processor(s) architecture blocks: (a) single core; (b) dual-core; and (c) N-core.

should be handled by the processor alone using priorities defined. This process suffers from a delay in execution in case of urgent need of speed in execution. Whereas in the multi-core (either dual or more, here symmetrical multi-core [3] architecture is considered) the execution runs on more than one core, but the instructions are all normal instructions in a program. Thus, the execution is done with the help of more than one processor which ultimately reduces the execution time. Figure 4.3 shows the IBM machine's single and dual-core block diagrams for HPC. The thrashing of the cache problem is overcome in this design with a memory loading compromise [4].

Table 4.1 shows some of the Intel's processor evolution concerning the number of processing cores from inception. This gives an idea that how the processors were developed with the time and processing features were added to make the processing easier even for a basic user (programer). Like the introduction of co-processors to support the main processor (Intel's 8088 to support 8086 as a math co-processor).

4.3   THE NEED FOR MULTI-CORE ARCHITECTURE

In the previous Section 4.2, we discussed the basic features and functionalities that a single and multi-core processors can do. A single-core processor utilizes all the resources available onboard, like RAM, I/O, and allows the peripherals to have access to these on basis of their need and priority of that particular peripheral (like DMA, USART, etc.), by its own. But in the case of multi-core as more than one processor is there on the chip hence an additional controller is required to bridge the usage without a deadlock and considering the design scaling [2]. This is the basic idea of multi-core architecture design.

In single-core systems, if a particular task has to be done in the specified time, which is the over capability to that processor due to the clock and processing bit capacity constraints it will surely fail to fulfill the task in the required time. Like processing of a big-data, video processing, satellite image processing, and medical image processing. If we can share the task between two or more processing elements (PEs) in parallel, then the results can be achieved in a specific time demand. The data that produced every day by individuals, organizations, and industries are nearly a huge one that cannot be handled by a single-core efficiently. With the help of multiple CPUs on a single chip, parallel computing can be done with

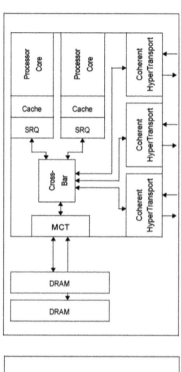

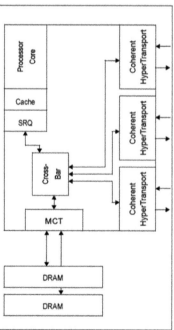

FIGURE 4.3 Block diagrams of IBM single and dual cores: (a) single-core architecture; and (b) dual-core architecture.

TABLE 4.1 The Evolution of Few Intel's Processors from Inception (Major Features Covered)

Name	Year	Processing Capacity and Type	Clock	Number of Transistors	Technology
4004	1971	4-bit Single-core	108–740 kHz	2,300	10 microns
8008	1972	8-bit Single-core	200–800 kHz	3,500	10 microns
8080	1974	8-bit Single-core	2 MHz	4,500	6 microns
8085	1976	8-bit Single-core	3 MHz	6,500	3 microns
8086	1978	16-bit Single-core	5–10 MHz	29,000	3 microns
80286	1982	16-bit Single-core	6–25 MHz	134,000	1.5 microns
80386	1985	32-bit Single-core	16–33 MHz	275,000	1.5 microns
80486	1989	32-bit Single-core	25–50 MHz	1.2 million	1 micron
Pentium	1993	32-bit Single-core	60 or 66 MHz	3.1 million	0.8 microns
Pentium Pro	1995	32-bit Single-core	200 MHz	5.5 million	0.35 microns
Pentium-II	1997	32-bit Single-core	300 MHz	7.5 million	0.25 microns
Celeron	1998	32-bit Single-core	266–300 MHz	7.5 million	0.25 microns
Pentium-III	1999	32-bit Single-core	500 MHz	9.5 million	0.25 microns
Pentium-4	2000	32-bit Single-core	1.5 GHz	42 million	0.18 microns
Xeon	2001	32-bit Single-core	1.7 GHz	42 million	0.18 microns
Pentium-M	2003	32-bit Single-core	1.7 GHz	55 million	130 nm
Core Solo and Duo	2006	32-bit 1 or 2 cores	1.06–2.33 GHz	151 million	65 nm
Core-2	2006	64-bit 1, 2 or 4 cores	2.66 GHz	291 million	65 nm
Atom	2008	32-bit 1–8 cores	1.86 GHz	47 million	45 nm
Core i7	2008	64-bit 4–10 cores	2.66–3.2 GHz	731 million	45 nm
Core i5	2009	64-bit 2 or 4 cores	2.66 GHz	774 million	45 nm
Core i3	2010	64-bit 2 cores	2.93–3.07 GHz	382 million	32 nm
Core i9	2017	64-bit 10–18 cores	2.6–3.3 GHz	–	14 nm

which the efficiency and speed are boosted up. Hence the requirement of a multi-core processor architectures came into the picture of modern processor manufacturing. In this section, we explore a few design methods, architectures, and challenges in the design of multicore processor chip architectures [1]. The manufacturing technologies had a step forward from single-core to multi-core and further to multi-systems design on a single chip (system on chip (SoC)). As the data handling is increasing from megabytes to gigabytes further terabytes and on, the system performance is measured on MIPS (million instructions per second) to mega and teraFLOPS (floating-point operations per second (FLOPS)). All modern manufacturers of the multi-core processors and SoCs are widely using system integration techniques and validating their design.

4.3.1 SoC ARCHITECTURE

The SoC is a single chip in which a core (CPU) or multiple-cores (CPUs) along with most of the required peripherals like memory, I/O, control modules are connected with busses and manufactured into a single chip. Figure 4.4(a) shows the basic block diagram of an SoC, and Figure 4.4(b) shows the design flow of SoC that defines hardware and software codesign.

A processor with high power-efficiency and high-speed hardware peripherals are required to produce the best possible design architecture which compromises the performance and power consumption. Since dynamic power emission increases with computational power exponentially [5]. System-on-chip technologies Inc., based H.264, MPEG-2, and H.265 CODEC IP Cores, CODEC Chipsets, and CODEC SOM Modules are recently developed for medical image/video processing, airborne video systems, video surveillance, high-definition video, satellite video, and video broadcasting applications which used an Intel's multi-core processors for their design [6, 7]. In the 2nd gen EPYC of AMD, it has 64-core/128 threads and 8-channel DDR4 with ECC up to 3,200 MHz. It uses an SoC integration methodology in this design. AMD is further planning an approximate of 1.5 ExaFLOPS higher performance HPC soon [8].

4.3.2 SoC DESIGN CHALLENGES AND SOLUTIONS

With the increasing number of integrating cores on a single chip, the efficiency of communication on the chip has become a major design challenge. Traditional busses are not meeting the expectations of speed and performance since they are limited in sharing. Shared-global bus gives a cost-efficient solution with its simple structure but it is efficiency limited to few tens of cores. This is because of its non-scalable wire delay with the technology shrinking and greater power consumption in long wiring.

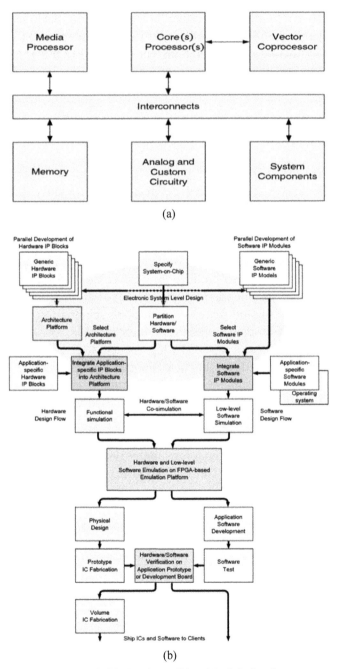

(a)

(b)

FIGURE 4.4   (a) SoC generic blocks; (b) traditional SoC design flow.

Figure 4.5 shows the delay in logic gates for various technologies using local and global wires as a communication interface.

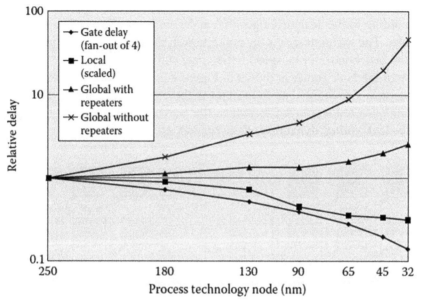

FIGURE 4.5   Local wire, global wire delays for logic gates at various technologies.
Source: Data referred form: ITRS's technical-reports.

4.3.2.1   NETWORK ON CHIP (NoC)

Network on chip (NoC) is one of the best solutions for the communication problems of SoC. Most of the communication problems are addressed with the 2D and 3D NoC designs. An NoC provides communication among the different PE or IP-cores that are integrated on a single chip SoC.

4.3.2.2   NETWORK ON CHIP (NoC) ARCHITECTURE

The NoC architecture with a 3 × 3 mesh topology is shown in Figure 4.6(a) and (b) shows a sample routing from source to destination. The major design elements in an NoC are: (i) network interface (NI); (ii) router design; and (iii) switching and choosing network topology. Choosing an efficient network topology is essential in the NoC design according to the structure or architecture of the SoC based on average distance. Network topologies like spidergon, octagon, mesh, star, torus, folded torus, etc., in either 2D or 3D. The NI is responsible for scaling the data into packets according to the designed algorithm and sending the data to the specified router. The router design is the major task in the entire NoC design, which is the benchmark for its speed, efficiency, and performance. The architecture of the NoC router is shown in Figure 4.7. A network crossbar switch is used to switch the data from input to the output of the router. Different techniques are employed to reduce the power consumption at this level like clock gating, dynamic voltage scaling, etc.

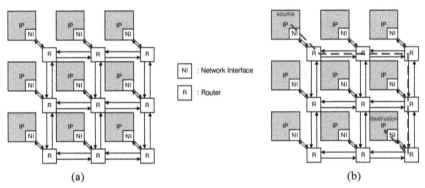

FIGURE 4.6   (a) NoC architecture; and (b) data communication between different cores of SoC using NoC as a communication interface.

To design an efficient multi-core processor architecture for high-performance computing one should follow the methods, techniques, and design specifications properly.

4.4   MULTICORE ARCHITECTURE AND ITS INTERACTION WITH OPERATING SYSTEM (OS)

In the previous section, we have seen the hardware level abstractions for a high-performance computing multicore architecture. It is also important to consider that, the developed hardware efficiency in interacting with the operating system (OS) is how much. The major challenge in multi-core computing is the unrivaled scale of demand on resources with huge uncertainty in the demand. The OS services should be elastic and scalable to the change of the demand. Hence shifting of resources between multiple systems services concerning changing loads is desirable to achieve high performance [9]. OS generally contains virtual-machine monitors or hypervisors which manage the various hardware platforms on a computer. Hence the hardware designer of multi-core processors and its supporting blocks has to look over the basic standard framework programming (OS) plan decisiveness to assist multi-cores and to manage them efficiently with on-chip hardware-assets that are shared with multiple-cores. No doubt the added hardware features of multi-core will enhance the performance of the software, but the quality of service (QoS) should not be a compromise [10]. Because adding additional hardware causes extra latency, extra power consumption, etc. Hence the new design architectures should not be a cutter of the performance of an OS.

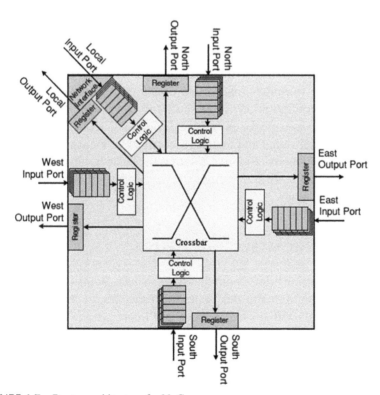

FIGURE 4.7 Router-architecture for NoC.

Some scientific researches and developments are rising to mitigate the hardware and software co-design issues. Kyle J. Nesbit et al. defined two extraneous procedures for virtualization: temporal and spatial [11]. There is a benchmark in the minimum performance which the VPMs (virtual-private-machines) will achieve irrespective of the number of tasks in the multi-core system. Fred A. Bower et al. defined the importance of an OS in dealing with heterogeneous multicore architectures at runtime. If not, a considerable amount of power loss and efficiency loss will be unavoidable [12]. Stijn Eyerman et al. refined the multiprogramming work-loads. According to this approach, the system performance should be done in the top to down way, starting from program-turnaround-time and system-throughput [13]. According to many research findings, it is understood that both hardware developers of multicore and software developers like OS designers should consider the compatibility issues and should design or develop dynamically adaptable environments.

Few Embedded engineering designers had developed a new OS with a powerful kernel to communicate with multi-core architectures. Microsar-OS allows separate executions of all the application-software on multi-processor-cores (Vector Informatik, Stuttgart, Germany). With the help of coordinated-access and synchronization methods, the OS will communicate with the resources that are in common. Microsar-OS Multicore kernel's static configuration yields short and fastest results of execution. Hence it is more suitable for embedded designs and applications like DSPs and DIPs (Digital signal and image processing) in domestic, medical, and industrial environments. Figure 4.8 shows the basic building blocks of the Micorsar-OS Multicore architecture and the and its multi-task handling for different applications.

4.5 INSTRUCTION-LEVEL PARALLELISM

The previous section briefs about the issues that are facing by the hardware-software co-designers and the issues with the OS on multi-core chips. This section deals with the core idea of multi-core processing systems, the instruction-level parallelism. Basic single-core processors will execute all the instructions in serial and the total execution time 't' of CPU is:

t = Instruction count × CPI × clock cycle time (or)
  = Instruction count × CPI/clock rate

where; CPI is the clocks per instruction.

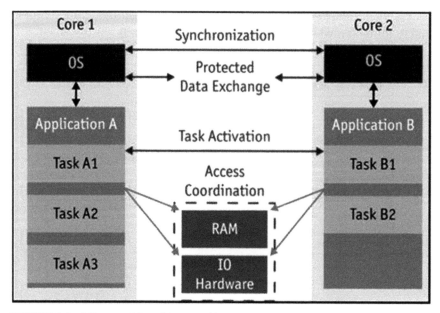

FIGURE 4.8 Microsar-OS multicore architecture.

If the task is larger than the execution time will increase in single-core systems. Hence in the design of multicore processor architectures, the instruction-level parallelism (ILP) is inevitable that can handle multi-threads parallelly to speed the execution process and reduce the execution time compared with single CPU systems. ILP is a measure of how many independent instructions in a computer can execute parallelly. The classification of ILP-architecture is in three configurations [14]:

1. **Sequential-Architectures:** In this architecture, parallelism is ambiguity as no clear information is there from the program regarding parallelism. Here superscalar-processors handle ILP.

2. **Dependent-Architectures:** In this architecture, parallelism has no ambiguity as clear information is there from the program regarding parallelism. Here dataflow-processors handle ILP.

3. **Independent-Architectures:** In this architecture, parallelism has no ambiguity as clear information is there from the program regarding parallelism and operations that are independent of each other. Here Very long instruction word (VILW)-processors handle ILP.

Figure 4.9 shows the division of these three procedures allying compiler and runtime hardware for the above three architectures.

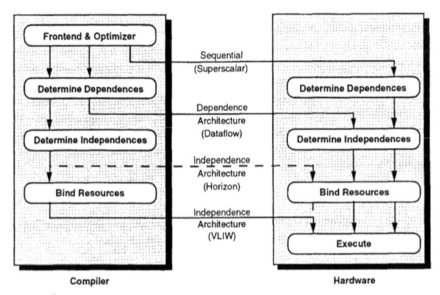

FIGURE 4.9   Handling ILP tasks allying compiler and hardware for ILP-architectures.

ILP in multi-core architectures has a different level of abstraction. Different processors take radically different approaches, for example, consider CPUs and GPUs in a multi-core:

- **CPUs: Instruction-Level Parallelism:**
  - Implicit;
  - Fine-grain.
- **GPUs: Thread-and Data-Level Parallelism:**
  - Explicit;
  - Coarse-grain.

In the case of CPUs of a multi-core, the hardware dynamically schedules the instructions for parallelism. But it is expensive and complexity and cost increase with the integration of more cores, even a few dozens. Anyhow it is relatively easy to write fast software. Similarly in the case of GPUs software makes the parallelism explicitly. It is a simple and cost-effective few hundreds of crores can be integrated. But it is hard to write fast software. Nowadays CPUs exploit ILP to speed up straight-line code with key ideas of:

1. **Pipelining and Superscalar:** Work on multiple instructions at once.
2. **Out-of-Order Execution:** Dynamically schedule instructions whenever they are "ready."
3. **Speculation:** Guess what the program will do next to discover more independent work, "rolling back" incorrect guesses.

Parallelism requires independent work, but many instructions are dependent, and many possible hazards like data-hazards, control-hazards with speculation limit parallelism. ILP works great and effective in the hardware perspective, but it is hard to scale in multicore architectures since parallel software is very hard.

4.6 THREAD-LEVEL PARALLELISM (TLP)

A 'thread of execution' simply shortly in computer science terminology 'a thread,' is a path in a program, split self into two or more concurrently or faux-concurrently running tasks. The 'process' and 'threads' differ from one OS to another one, even in general a thread (like in a document user typing is one thread, spell check, printing, etc., are other threads) is a subset of a process (like playing music, surfing the web, running program: maybe a print).

4.6.1 MULTITHREADING

Most of the time ILP fails with resulting relatively very low parallel executions when more execution units were added to superscalar and/or Out of Order (OoO) process. Alternatively, a new technique called 'multithreading' will support the processor with multiple executions-units running, even though the ILP is stalled in stand by for memory or low for program.

Thread-level parallelism (TLP) also termed as task-parallelism or function-parallelism is a structure of parallel-computing. TLP technique is to assign the execution of 'processes,' 'threads' over various parallel-processor nodes. TLP plays a vital role in the executions of programs in multi-processor or multi-core systems, it succeeds once if each processor executes a non-identical process or non-identical thread upon similar or dissimilar data. In TLP the threads may execute similar or dissimilar program-codes. Non-identical execution-threads can communicate among them while working to pass data. TLP is more cost-effective over ILP. Figure 4.10 illustrates the TLP pictorially.

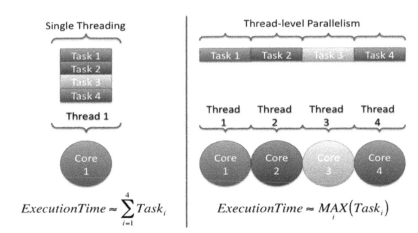

FIGURE 4.10   Single thread vs. TLP.

The main strategies of TLP are SMT (Simultaneous Multi-Threading) and CMP (Chip Multi-Processing). In SMT multiple threads execute simultaneously and in CMP Multiple cores are on the same die where each core can probably execute one type of execution. In general multicore-processor packs multiple-processors on the same die which will share the main-memory and occasionally bus, and second-level-cache. Every processor is associated with its own first-level-cache (L1 cache). The TLP or embedded-tasks allowed by the processors which are independently executing in the multi-core processors case [15].

4.7   GENERAL CONTEXT: MULTIPROCESSORS

The above sections have described multi-core design ideas in processor manufacturing in contrast to both hardware and software co-design. This section focusses on the classes of the modern multi-core processors to some extent. The novel processor designs come with a few hundreds of millions of transistors to some trillion number of transistors (refer Table 4.1). With the help of these, we can increase the Pipeline-depths or a good number of execution-units can be appended to superscalar processors that might raise a little considerable amount of performance hike, which is not a power-efficient technique. For one and half decades the manufacturers of computer processors are developing multiple copies of the same processors with the name cores [16]. The main three types of classes in multi-processor consists of: (i) homogeneous (symmetric) multi-processors; (ii) heterogeneous (asymmetric) multi-processors; and (iii) clustered-multi-processors.

4.7.1 SYMMETRIC MULTIPROCESSORS

In this design method two or more processors that are the same in design architecture (like cloning or copying) are sharing the available main memory. If the processors are integrated on the same chip, then they are called multi-core. Figure 4.11 shows the basic architecture. They help to run multiple numbers of threads of a process in parallel or to speed up one thread in a process, and it a challenging task. The difficulty is that the programer has to assign each thread to the available processor cores, and communication among them is also a challenge. Utilizing a large number of the cores available is not an easier one for the programers. Besides, they are much simple in designing and replication is easier. Further running a program is easier in these symmetric multiprocessors as each processor gives the same execution speeds. These are good for large data centers.

4.7.2 HETEROGENEOUS MULTIPROCESSORS

Adding the same symmetric cores is always not giving an improvement in performance every-time. For a few years, user's demand for execution speeds increasing drastically for different applications that are running parallelly (like mobile phone apps). General processors of symmetric core (even dual-core, quad-core) give an average performance and power inefficient, thus continuing of adding more cores will saturate the performance at one point and may be decline the performance. Hence Heterogeneous Multiprocessors systems are aimed to address these problems by introducing some specialized hardware in the system. So, depending on performance and power requirements, every application can utilize a particular hardware resource. These systems will come with a variety of microarchitectures with various power trade-offs. Figure 4.12 shows the basic architecture. Now a day mobile phones even coming with graphic processors, wireless radio, DSPs, and DIPs, etc. The complexity in designing in both hardware and software are a few limitations of this approach.

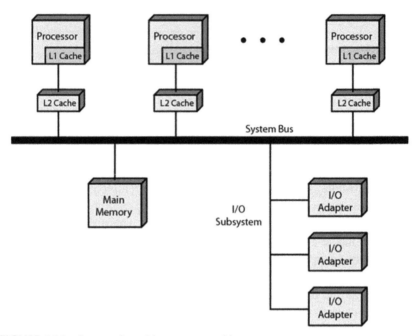

FIGURE 4.11   Symmetric multiprocessors architecture.

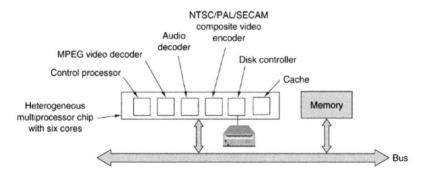

FIGURE 4.12   Heterogeneous multiprocessors architecture.

4.7.3   CLUSTERED-MULTI-PROCESSORS

The main advantage here is having local memory for each cluster of multi-processors on its own. They share common resources like power and cooling equipment (like in data centers). Networking of autonomous computes for a grater job handling will also come under a cluster.

4.8 MULTIPROCESSOR MEMORY TYPES

Every processor requires system memory (RAM, ROM, Chache, etc.), for data processing or execution. How effectively it can have access that effective is the performance. Since in the context of multiprocessors, many cores of homogeneous or heterogeneous are in a single die, the memory accessing is not an easier job [17]. Dedicated methods and algorithms are required to be implemented to avoid memory congestions or deadlocks which affect the system performance. Broadly used memory access types in multiprocessor chips are: (i) shared memory multiprocessors; and (ii) distributed memory multiprocessors.

4.8.1 SHARED-MEMORY MULTIPROCESSORS

In the shared memory multiprocessor architecture, all the PEs or processors on the core have straight access to all the main memory. When cache-coherence is considered, this design leads to more complicated as all caches need to replicate the change in one cache. Figure 4.13 shows the system level block diagram of a shared memory concept of multiprocessors, (a) shows the basic shared memory multiprocessor (SMM) in which access time of all the cores are uncertain and depending on the load of the processor and (b) shows symmetric shared-memory multiprocessor (SMP) provides uniform access time for all cores.

Most of the modern OSs with a defined scheduler support SMP. Load balancing is the major concern in the multiprocessor architectures.

4.8.2 DISTRIBUTED MEMORY MULTIPROCESSORS

In the context of distributed multiprocessor memory, all the cores have their own private memory for execution of tasks locally with the limited resources for faster executions. Figure 4.14 shows the distributed memory concept.

Some methods like uniform memory access (UMA) in which all cores will have the same access time, non-uniform memory access (NUMA) in which all cores will have the access time depending on data position in local-memory, and cache only memory access (COMA) in which all cores' data can migrate and/or copied on various memory-banks of the main memory, are used for enhancing the memory usage in multiprocessors.

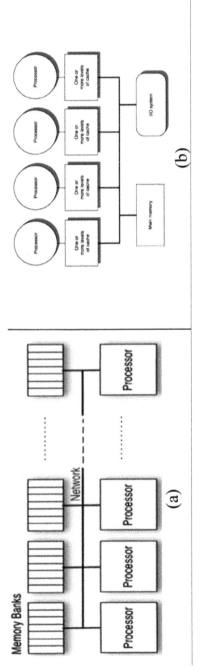

FIGURE 4.13   (a) SMM architecture; and (b) shows symmetric shared-memory multiprocessor (SMP) architecture.

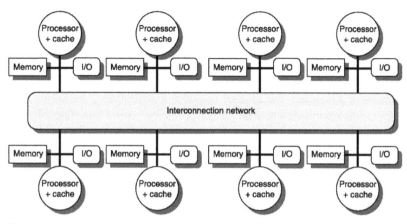

FIGURE 4.14   Distributed memory architecture in multiprocessors.

4.9   MULTICORE AND MULTITHREADING

In this section, we focus on market-level multicores and multithreading functions.

4.9.1   MULTI-CORES

According to the survey done by ESDJ, Truly et al. [18] proven that multi-processor manufacturing is heterogeneous by more than 50%. Figure 4.15 shows the result.

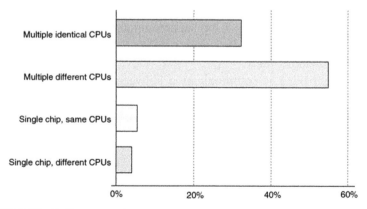

FIGURE 4.15   SoC processor types.

Source: Adapted from Ref. [19].

It is clear that the manufacturing and design of modern processors are in multi-core like the dual, quad, hexa, octa, deci, etc., since one and half decades. The usage of several cores decides the name, for example, if two processors are integrated on a single silicon chip it is a dual-core processor (processing capacity is depending on design, like 32, 64) vice versa. Figure 4.16 illustrates the idea of the core distribution. The major task is the CPU has to coordinate with all the cores for proper handling of tasks.

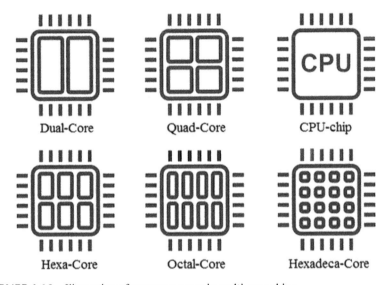

FIGURE 4.16   Illustration of processor cores in multi-core chips.

Figure 4.17 shows today's top supercomputers in the market and their corresponding number of cores used; it shows that the usage of processing cores is increasing widely.

The major thing is communication between the cores and resources without degrading the performance of the system. Figure 4.18 shows some of the parallel computing architectures to achieve the same.

4.9.2   MULTITHREAD

1. **Thread:** Process with own instructions and data, it may be a part of a parallel program of a multiprocessor or maybe an independent program.

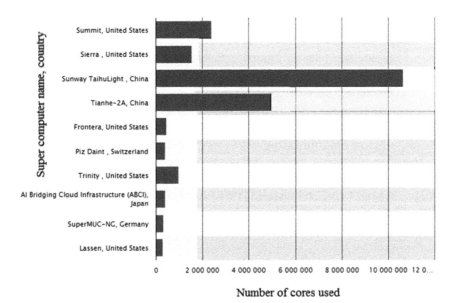

FIGURE 4.17 Illustration of processing core usage in supercomputers, ©Statista 2020.

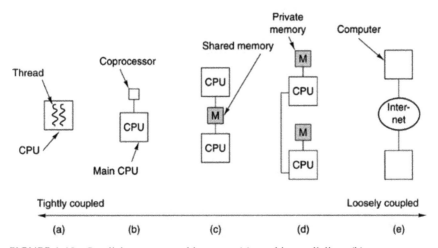

FIGURE 4.18 Parallel computer architectures: (a) on-chip parallelism; (b) a coprocessor level; (c) a multiprocessor level; (d) a multicomputer level; and (e) a grid level.

2. **Multithread:** Multiple threads to share the functional units of one processor via overlapping. The processor must duplicate the independent state of each thread, for example, a separate copy

of the register file, a separate core, and for running independent programs, a separate page table. In multithread execution thread switching, like an alternate thread for every clock should run. Because if a thread is stalled perhaps for a cache miss another thread can execute (cross-grain).

3. **Simultaneous Multithreading (SMT):** Insights that dynamically scheduled processor already has many hardware mechanisms to support multithreading.

4.10 SIMULTANEOUS MULTITHREADING (A TECHNIQUE COMPLEMENTARY TO MULTI-CORE)

In the previous section, we have covered the basic forms of threading and their definitions and functions. In coordination with the increasing cores per silicon even in general processing units, SMT has its mark in task executions. Increasing the number of cores to achieve high performance is not a good solution for an ever due to complex circuits (more number of transistors) and power consumption problems. The overall throughput of the system will increase with multithreading with a compromise of less increase of ILP. Thus 'Simultaneous Multithreading' evolved as an alternative to complementary multi-core architectures problems. Table 4.2 shows the comparison of different multicore processor's methods of switching between threads-per core. Most of them are using SMT (Intel calls it as Hyperthreading-Technology HTT [19]) which sends precoded instructions from only that belongs to on-chip-threads per cycle. Dean M et al. stated and showed that the usage of SMT gives better results in increasing the chip processing performance over single-chip multiprocessing [20].

4.11 THE BENEFITS OF MULTI-CORE IMAGE PROCESSING

The above sections are given a good idea on some concepts of processing, parallel processing, multiprocessing techniques with system level, core level multithreading. With all, that one can understand that a multicore system with the use of multithreading can handle a good amount (big) of data with usually less amount of processing time than usual. This is the most important thing required in the field of image processing. Because image processing hardware/software environments require a lot of PEs

TABLE 4.2 Comparison of a Few Commercial Multicore CPUs for Multithreading [19]

Vendor	Product	Cores	Threads		Clock (GHz)	Power	Special Features	On-Chip Interconnect	L2 Size per Chip (Mbytes)	L2 Allocation	L3 Size (Mbytes)
			Per Core on-Chip/ Executing	Switching Approach							
Intel	Itanium (9000 series)	2	2/1	Blocked+	1.4–1.66	75–104	VLIW, IR 6	Direct pathways	I: 2 D: 0.5	Private	4, 6, 9, 12 per core, private
Intel	Xeon (7400 series)	4, 6	2/2	SMT	2.13–2.66	50–130	IR 4, dynamic PM	On-chip bus	6, 9	Shared	8, 12, 16
Intel	Core i7	4	2/2	SMT	2.66–3.33	130	Triple channel IMC	Crossbar	1	Private	8
IBM	Power5	2	2/2	SMT	1.5–1.9	Unpublished	IR 5, IMC	Crossbar	1.875	Shared	36 off-chip
IBM	Power6	2	2/2	SMT	4.7–5	Unpublished	IR 7, 1 decimal, 2 binary FPUs per core	On-chip bus	8	Private	32 off-chip
IBM	Cell BE, PPE	1	2/2	SMT	3.2	110+	General purpose	Ring bus	0.5	N/A	N/A
IBM	Cell BE, SPE	8	N/A	N/A	3.2	110+	Simplified for SIMD support	Ring bus	N/A	N/A	N/A
Sun	UltraSPARC T1	4, 6, 8	4/1	Interleaved++	1.0–1.2	72–79	1 FPU per chip, IMC	Crossbar	3	Shared	N/A
Sun	UltraSPARC T2	4, 6, 8	8/2	Parallel interleaved Δ	1.0–1.6	95–123	IMC and INC, crypto unit (per core), SOC, 1 FPU per core	Crossbar	4	Shared	N/A
Sun	UltraSPARC IV+	2	N/A	N/A	1.5–2.1	90	IR 4, IMC	On-chip bus	2	Shared	32 off-chip

TABLE 4.2 (Continued)

Vendor	Product	Cores	Threads Per Core on-Chip/Executing	Threads Switching Approach	Clock (GHz)	Power	Special Features	On-Chip Interconnect	L2 Size per Chip Mbytes	L2 Allocation	L3 Size Mbytes
Sun	Sun/Fujitsu SPARC64 VII	4	2/2	SMT	2.5	135	IR 4, hardware barrier	On-chip bus	6	Shared	N/A
Sun	Rock	16	2/2	SMT	2.1	250	4-core clusters, IR 4, aggressive speculation, HTM, 2 FPUs per cluster	Direct pathways/crossbar (among clusters)	2	Shared	16 off-chip
Specialized	Tilera TILE 64	64	N/A	N/A	0.5–0.9	15–22	Simple cores, no FPUs	Multilink mesh	4†	Shared†	N/A
Specialized	ARM Cortex-A9 MPCore	2, 4	N/A	N/A	1	<1	Ultrasmall, SOC, ultra-low-power	Multilevel bus	2	Shared	N/A
Specialized	ATI RV770†‡7	10	>1,000‡/10	Interleaved	0.75	160	Simplified for SIMD, 80 FPUs per core	Crossbar	>256 Kbytes‡	Shared	N/A
Specialized	Nvidia GT200††8	30	1,024/8–16	Interleaved	1.295	236	Simplified for SIMD, 10 FPUs per core	Crossbar	256 Kbytes	Shared	N/A

with good handling algorithms. Hence Multicore systems reliable solutions for image processing and multimedia applications.

One of the exhaustive tasks in complex computations is 'Image processing' consumes a huge amount of resources like CPU and memory. Parallelism techniques with traditional multicore processors like a dual, quad, etc., will help to some extent but, suffers greater power consumption. Thus, heterogeneous multicore processors like image processing dedicated hardware (GPUs, DIPs, DSPs) can solve the processing issues.

Shams et al. [21] worked on the state-of-the-art techniques for good medical image processing with the use help of massive-multiprocessing (MPP), High-performance computing (HPC) architectures with symmetric multiprocessing techniques proven that the use of multicore architectures benefit in enhancing the speed in medical image processing.

Saxena [22] worked on reliable and accurate methods to process the medical images using suitable artificial intelligence (AI) techniques on multicore architectures using parallel computing found that the usage of multicore architectures in image processing reduces the processing times.

4.12 EVALUATION OF MULTI-CORE ARCHITECTURES FOR IMAGE PROCESSING ALGORITHMS

The greater capacity of computations and programming powers of multicore architectures yields greater hope for advancements in computer-vision, image-processing, medical, and industrial image processing, and surveillance applications with less false positives. Classification of image processing algorithms is at: (i) low-level; (ii) intermediate-level; and (iii) high-level [23–25].

1. **Low-Level Image Processing:** It works on total image to find a single value. The image data will be spatially localized with minimum operations to achieve parallelism. It mostly uses single instruction, multiple data (SIMD) architectures. Histogram generation, shaping, smoothing, etc., done at this level.

2. **Intermediate-Level Processing:** These are compact and get like-list form inputs. These works on segments. Motion-analysis, Object-labeling, Hough-transform are done at this level.

3. **High-Level Processing:** These are the symbolic process that works on data structures as arguments and returns. Involves huge sets of complex operations and hardware resources.

Based on the requirements of the developer, one has to choose a proper level and suitable algorithms and make use of advanced hardware like Multiprocessors, Chip-Multiprocessors, MPSoC, to yield good results. Figure 4.19 shows a typical 'Ultrasound' system contains major hardware and processing blocks in medical imaging. Figure 4.20 shows the flow of high-quality image processing.

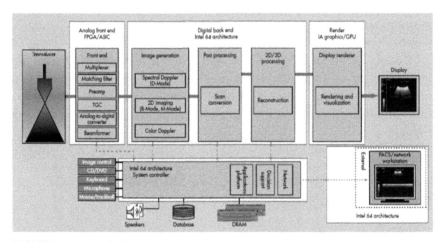

FIGURE 4.19   Typical ultrasound system.

Source: Courtesy of Intel.

4.13   CONCLUSION

The summary of all the sections is time, speed, cost minimizations with no compromise on performance and quality. Multicore architectures and parallel processing techniques will fulfill the same with the best QoS and user satisfaction. The mobile manufacturing field is a live example of the same. Day to day the PEs and processing applications are increasing with no compromise on the user data as megapixels of cameras are getting integrated. Augmented reality and Virtual reality are pocket friendly and handy nowadays with the inclusion of Multi-core Architectures in mobile manufacturing, besides heat dissipation and power consumption.

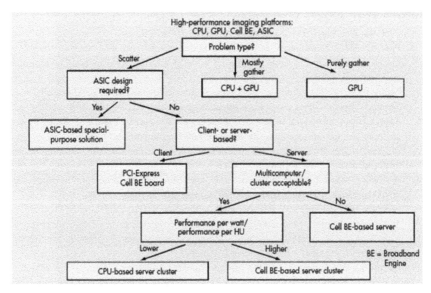

FIGURE 4.20   High-quality imaging flow.

Source: Courtesy of Mercury Computer Systems.

KEYWORDS

- **network interface**
- **network on chip**
- **operating system**
- **processing elements**
- **quality of service**
- **system on chip**

REFERENCES

1. Geer, D., (2005). Chip makers turn to multicore processors. In: *Computer* (Vol. 38, No. 5, pp. 11–13). doi: 10.1109/MC.2005.160.
2. Esmaeilzadeh, H., Blem, E., Amant, R. S., Sankaralingam, K., & Burger, D., (2011). Dark silicon and the end of multicore scaling. In: *2011 38th Annual International Symposium on Computer Architecture (ISCA)* (pp. 365–376). San Jose, CA.

3. Hill, M. D., & Marty, M. R., (2008). Amdahl's law in the multicore Era. In: *Computer* (Vol. 41, No. 7, pp. 33–38). doi: 10.1109/MC.2008.209.

4. Pase, D. M., & Matthew, A. E., (2005). *A Comparison of Single-Core and Dual-Core Opteron Processor Performance for HPC*. IBM xSeries Performance Development and Analysis 3039.

5. Najm, F. N., (1994). 'A survey of power estimation techniques in VLSI circuits. *IEEE Transactions on Very Large-Scale Integrations Systems, 2*(4), 446–455.

6. Sources from the official website of "*System-on-Chip Technologies Inc.*" https://www.soctechnologies.com/modules (accessed on 30 September 2021).

7. Intel's official: https://www.intel.in/content/www/in/en/products/programmable/soc.html (accessed on 30 September 2021).

8. AMD's: https://amdhub.in/server.html and https://www.amd.com/en/processors/epyc-world-records (accessed on 30 September 2021).

9. Wentzlaff, D., et al., (2010). An operating system for multicore and clouds: Mechanisms and implementation. *Proceedings of the 1st ACM Symposium on Cloud Computing.*

10. Cho, S., Li, T., & Mutlu, O., (2008). Interaction of many-core computer architecture and operating systems. *IEEE Micro, 28*(3), 2–5.

11. Nesbit, K., & Laudon, J., (2007). *Virtual Private Machines: A Resource Abstraction.*

12. Bower, F., Sorin, D., & Cox, L., (2008). The impact of dynamically heterogeneous multicore processors on thread scheduling. *Micro IEEE, 28*, 17–25. 10.1109/MM.2008.46.

13. Eyerman, S., & Eeckhout, L., (2008). System-level performance metrics for multi-program workloads. *Micro IEEE, 28*, 42–53. 10.1109/MM.2008.44.

14. Rau, B. R., & Joseph, A. F., (1993). Instruction-level parallel processing: History, overview, and perspective. *Instruction-Level Parallelism* (pp. 9–50). Springer, Boston, MA.

15. David, L., (2013). Chapter 14 - high-performance business intelligence. In: David, L., (ed.), *MK Series on Business Intelligence, Business Intelligence* (2nd edn., pp. 211–235). Morgan Kaufmann. ISBN 9780123858894. https://doi.org/10.1016/B978-0-12-385889-4.00014-4.

16. Sarah, H., & David, H., (2015). *Chapter 7: Digital Design and Computer Architecture* (2nd edn., pp. 385–477). doi: 10.1016/B978-0-12-800056-4.00007-8.

17. Sterling, T., (2018). *Chapter 2: High Performance Computing || HPC Architecture 1.* doi: 10.1016/B978-0-12-420158-3.00002-2.

18. Turley, J., (2005). Survey says: Software tools more important than chips. *Embedded Systems Design Journal.*

19. Marr, D., et al., (2002). Hyper-threading technology architecture and microarchitecture. *Intel. Technology J., 6*(1), 4–15.

20. Dean, M. T., Susan, J. E., & Henry, M. L., (1995). Simultaneous multithreading: Maximizing on-chip parallelism. *Proceedings of the 22nd Annual International Symposium on Computer Architecture.* Santa Margherita Ligure, Italy.

21. Shams, R., et al., (2010). A survey of medical image registration on multicore and the GPU. *IEEE Signal Processing Magazine, 27*(2), 50–60.

22. Saxena, S., (2013). Image processing tasks using parallel computing in multi core architecture and its applications in medical imaging. *International Journal of Advanced Research in Computer and Communication Engineering, 2*. 10.17148/IJARCCE.2013.2420.

23. Komen, E. R., (1990). *Low-Level Image Processing Architectures*. ISBN 90-9003713-6.

24. Saxena, S., Sharma, N., & Sharma, S., (2015). Parallel computing in genetic algorithm (GA) with the parallel solution of n queen's problem based on GA in multicore architecture. *International Journal of Applied Engineering and Research, 10*(17), 37707–37716.

25. Saxena, S., Sharma, N., & Sharma, S., (2016). Parallel image processing techniques, benefits and limitations. *Research Journal of Applied Science, Engineering & Technology, 12*(2), 223–238.

CHAPTER 5

MACHINE LEARNING APPLICATIONS IN MEDICAL IMAGE PROCESSING

TANMAY NATH and MARTIN A. LINDQUIST

Department of Biostatistics, Bloomberg School of Public Health, Johns Hopkins University, Baltimore, Maryland, USA, E-mail: tnath3@jhu.edu (T. Nath

ABSTRACT

The term "learning" relate to a broad range of processes that can be defined as "knowledge or skill acquired by instruction or study." Psychologists study learning in both humans and animals, but in this chapter, the authors focused on learning in machines and how this process has evolved during the last 60 years and found important applications in medical image processing. The author discussed how computational models built to understand human learning have led to two main categories of machine learning: supervised learning and unsupervised learning. They also presented the concepts of deep learning and how it is beginning to play a vital role in medical image processing. To help researchers interested in applying machine learning to medical imaging data, they have provided information about different data formats and available resources to analyze these data. The chapter concludes with a discussion of how machine learning is expected to continue to play an important role in medical image processing and, combined with a doctor's experience, will help improve medical outcomes.

5.1 INTRODUCTION

In a daily life many human activities like reading, writing, and driving, demand "intelligence." Recently, significant research effort has focused on making machines perform similar tasks. When a machine application is capable of learning a particular skill, it is termed "artificial intelligence" (AI) [1]. Some examples of AI include autonomous cars, mortality prediction, and understanding human speech.

The field of machine learning (ML) is complementary to the study of learning in humans and animals. Many techniques used in ML are in fact derived from computational models built to understand human learning process [2]. In contrast, the concepts and techniques used in ML are designed to help better understand aspects of biological learning [3].

ML is a class of techniques that allows computers to learn directly from a given set of examples. For example, in the task of identifying spam email, the machine's task is to learn to distinguish between spam and non-spam email and reliably assign the correct label ("spam" or "not spam") to every email that it receives. A traditional approach towards solving such problems relies on using rule-based learning [4]. For example, if the email contains certain keywords or statements, the machine is more likely to classify the email as spam. These approaches typically do not generalize well to new environments, and the rules need to be updated accordingly. In contrast, a ML algorithm learns a particular mathematical function describing the process by detecting patterns in a specific dataset. This function can be generalized to new environments making ML system more robust to changes in settings.

As a discipline ML lives at the intersection of computer science, statistics, and data science. It uses elements of each of these fields to process data in a way that can detect and learn from patterns, predict future activity, and make decisions. However, a question that naturally rises up is: "Why do we need machines to learn?" There are several reasons why ML is important and being used increasingly broadly. We have already mentioned how it might help us understand how animals and humans learn. However, there are other important engineering reasons [3] as well, including:

1. **Defining Tasks by Examples:** It is often difficult to provide a concise relationship between input and output pairs. However, a machine can extract relevant features from the input and find patterns that link the input and output pairs. For example, a machine can learn texture and other morphological features of an input cancer tissue image to identify whether or not it is benign; a classification task generally hard to perform by a human.

2. **Finding Patterns in Data:** ML methods can extract relationships buried deep inside the large amounts of data used to predict future trends. For example, in medical science, a data mining algorithm can be used to identify people at risk for diseases caused by environmental factors or genetic predisposition [5].

3. **Processing Large Amounts of Data:** Often, it is difficult for humans to encode the large amounts of information available about a certain task. For example, consider video annotation, where the goal is to identify a certain human or animal posture in a video frame. The large amount of information available within the entire video can be overwhelming for humans to process, and can be learned more reliably by a machine.

4. **Adapting to New Environments:** Environments change over time and machines need to be designed to adapt to these changes in order to respond appropriately. For example, an autonomous driving car needs to respond to situations that have not been encountered during its design.

5.1.1   ORIGIN AND EVOLUTION OF MACHINE LEARNING (ML)

The fundamental ideas that form the basis of the field of ML date back over 60 years ago when Alan Turing introduced the concept of the "Turing test" and coined the phrase "Can machine think?" [6]. In his seminal paper, he argues that human level intelligence could be demonstrated by an appropriately programed computer. Later, in 1952, Arthur Samuel demonstrated an early example of ML when he created the first game of checkers that could "learn" strategies as the game progressed.

The first artificial neural network (ANN) was developed in multiple stages. Its roots lie in the neurological work of Santiago Ramon Cajal who explored the structure of nervous tissues and demonstrated how neurons communicate with each other. These structures were combined with mathematical models to understand how neurons are able to make the computations needed to perform overt behaviors. Using knowledge from neuroscience, neurophysiologist Warren McCulloch and mathematician Walter Pitts developed the first model of a neural network in 1943 [7] and claimed that neurons have binary threshold activation functions. Although McCulloch-Pitts neurons were very simple and allowed only binary input and output, it gave Donald Hebb a platform to propose his revolutionary work in 1949. The Hebb's rule states "When an axon of cell A is near enough to excite a cell B and repeatedly or persistently takes part in firing it, some growth process or metabolic change takes place in one or both cells such that A's efficiency, as one of the cells firing B, is increased" [8]. This illustrates the fact that neural pathways are strengthened each time they are used which is conceptually similar to the way humans learn. This proposal became the fundamental operation necessary for learning and memory. Furthermore, Frank Rosenblatt combined knowledge obtained from McCulloch-Pitts neurons and the findings of Hebb to build the first perceptron in 1962 which later became instrumental in the formation of neural networks (NNs) [9].

In subsequent years, three major branches of ML emerged: symbolic learning; statistical methods; and NNs [10]. Symbolic learning algorithms learn new concepts by constructing a symbolic representation of objects in a class. For example, a symbolic learning system will construct a logic to distinguish between two objects based on attributes like size and color. Examples of symbolic learning systems include decision trees (DTs) [11] and inductive logic program [12]. Statistical methods, or pattern recognition methods, draw inspiration from statistics where one can use descriptive statistical methods to transform raw data and help a machine learn hidden patterns. Some of the examples include Naive Bayesian classifiers [13], K-nearest neighbors (KNNs), and discriminant analysis. Finally, NNs, described above, have advanced significantly since its inception as a perceptron. By adding multiple layers to the networks (e.g., multilayer feed-forward network with back-propagation learning), NNs are currently widely used for medical imaging applications. Moreover, with the advent of new technology and computational resources like graphical processing units (GPUs), Google tensor processing unit (TPU), cloud computing, a plethora of NNs have been developed, including AlexNet [14], ResNet [15], and U-Net [16]. See Table 5.1 for examples of common convolutional neural network architectures.

In further sections, we will review various ML methods and illustrate their application to medical imaging data.

5.2 MACHINE LEARNING (ML) ALGORITHMS

ML algorithms are explicitly designed to perform specific tasks by using examples and/or past experience. Roughly speaking, ML can be divided into two main categories: supervised learning and unsupervised learning. However, we will also describe deep learning (DL) in this chapter, which can be considered as either supervised or unsupervised depending on its application.

5.2.1 SUPERVISED MACHINE LEARNING (ML)

The most common category of ML algorithms is supervised learning, where the algorithm uses a training dataset to learn a mapping between input variables (or features) X and an output label y [17]. The algorithm uses this mapping to predict the output labels of an unseen new dataset (also known as the testing dataset). If $D_{training} = \{(x_1,y_1), (x_2,y_2)..., (x_n,y_n)\}$ represents a training dataset, where $x_i \in X$ is the i-th input data sample and y_i represents its corresponding output label, then the supervised learning algorithm estimates parameters θ of the mapping function $f: X \rightarrow Y$. In terms of medical image processing, x_i could be measurements from an image for subject i and y_i could be the potential outcome (e.g., benign, or non-benign cancer). Generally, there are two ways for the algorithm to learn the parameters θ of the mapping function [18]:

1. Maximize the score (ℓ) between the ground truth labels (y_{true}) and the predicted labels (y_{pred}) using the equation:

$$\max_{\theta} \frac{1}{n} \sum_i^n \ell\left(y_i, f\left(x_i, \theta\right)\right) \qquad (1)$$

One such score function is the hinge loss which maximizes the margin between the classes. A hinge loss can be expressed as $loss = maximum\ (1 - (y_{true} \times y_{pred}), 0)$ where the values of y_{true} are expected to be −1 or 1. It is commonly used in support vector machines (SVMs) [19]. Briefly, SVM is a supervised ML algorithm which makes the decision boundary separating two classes as wide as possible.

2. Minimize the loss ($\mathcal{L}$) over $D_{training}$ using the equation:

$$\underset{\theta}{argmin} \frac{1}{n} \sum_i^n \mathcal{L}(y_i, f(x_i, \theta)) \qquad (2)$$

There are many possible choices for such loss functions, including Huber loss and mean squared error loss. One of the most popular choices is cross-entropy loss, commonly used in convolutional neural network (see Section 5.2.3).

In the domain of medical imaging, the data may have more than 2 dimensions. For example, magnetic resonance images have 3 dimensions (x, y, and z spatial coordinates) and functional MRI images have 4 dimensions (x, y, z, and time). In these settings, the number of features is often greater than the number of subjects. This leads to a situation where there is a high chance that the algorithm overfits the data, resulting in very high accuracy on the training dataset that does not generalize to an unseen testing dataset. One way to guard against overfitting is cross-validation. Here the training dataset is split into k-folds. The algorithm trains on $k - 1$ folds while it is evaluated on the remaining fold. This procedure is repeated k times, with each fold being evaluated once, and is therefore referred to as k-fold cross-validation. The performance of each fold is analyzed and the parameters with the best performance are chosen. Additionally, this technique helps verify that the parameters behave consistently across each fold.

5.2.2   UNSUPERVISED MACHINE LEARNING (ML)

A second category of ML algorithms is unsupervised learning, which unlike their supervised learning counterparts do not require labels for training. Unsupervised learning is inspired by the mechanism behind human learning. A human brain receives substantial amounts of information throughout the day without any obvious reinforcement, yet is able to build "cognitive maps" [20] or "working models" [21] of the world around it.

For example, there are around 10^6 photoreceptors in each eye and their activity constantly changes with the environment. This provides necessary information regarding the presence of objects and their appearance. The activity patterns in these sensory neurons influence the structural and physiological properties of the synapse in the brain (neocortex region) [22]. Since, no instruction about the scene is provided during learning, the computational model of the synapse appears to learn in an unsupervised manner. That is, vast amounts of information are used to build features that provide knowledge about the patterns in the input stimuli using an unsupervised method [23].

Formally, an unsupervised ML algorithm is a technique which learns from the data without the need to specify any desired output. That is, unsupervised ML algorithms do not learn by mapping the input data to ground truth labels. Rather, the aim of an unsupervised ML algorithm is to discover hidden relationships in the training dataset which encode the structure of labels. However, it is possible that unsupervised learning techniques find patterns that do not necessarily have any semantic meaning in the context of the data domain. Therefore, one needs to be careful when drawing any conclusions. Nevertheless, unsupervised learning techniques open the door to the possibility of utilizing a large amount of unlabeled data to find patterns that can be eventually used in a supervised manner on smaller datasets. These techniques are extremely useful in the domain of medical imaging which tend to not have large labeled datasets. Two such techniques are semi-supervised and self-supervised ML.

5.2.2.1 SEMI-SUPERVISED MACHINE LEARNING (ML)

Semi-supervised ML algorithms are a set of learning techniques which use both labeled and unlabeled data to learn the mapping between the input dataset and its corresponding labels. The goal is to use the large set of unlabeled data with a small amount of supervision, in terms of labeled data, to learn a better prediction rule than can be obtained solely based on using labeled data [24]. It is also inspired by how humans implicitly combine labeled and unlabeled data to learn [25]. For instance, a child learns to categorize an object (e.g., an animal) to its corresponding correct label (i.e., a dog) with the help of a supervision (e.g., the child's parents or teacher pointing to the animal and saying "dog"). But more often children observe a lot of animals on their own and learn the correct "label" (i.e., correctly identify the animal) without receiving explicit supervision.

Practically, a semi-supervised ML technique makes the following assumptions regarding the data:

1. **Smoothness Assumption:** The algorithm assumes that nearby points are more likely to have the same output label.
2. **Cluster Assumption:** The data can be divided into discrete clusters, and points in the same cluster are more likely to share an output label.
3. **Manifold Assumption:** The data lie approximately on a manifold of much lower dimension than the input space. This assumption allows the use of distances and densities defined on the manifold.

Historically, the term 'semi-supervised' was first termed by Mertz et al. [26] in the context of a problem consisting of both labeled and unlabeled data. However, the concept of semi-supervised learning became popularized in the 1970s [27] in the context of estimating the Fisher linear discriminant rule using unlabeled dataset where each class conditional probability density was assumed to follow a Gaussian distribution with equal covariance matrix. Here, the likelihood of the model was maximized using the expectation-maximization algorithm (Figure 5.1) [28].

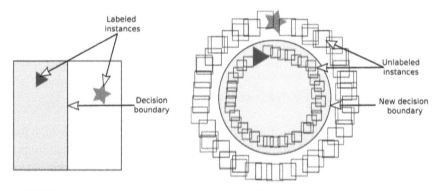

FIGURE 5.1   An example of semi-supervised learning.

Source: Inspired by Ref. [29].

Note: The image on left shows a linear decision boundary between the two labeled instances represented as a blue triangle and a green star. However, the image on right suggests that by adding additional unlabeled dataset represented as black squares, and using semi-supervised learning the particular geometric structure of marginal distribution suggests that the new decision boundary is circular which is more accurate.

The use of semi-supervised ML technique in medical imaging domain is discussed in Section 5.4.4.

5.2.2.2   SELF-SUPERVISED MACHINE LEARNING (ML)

Self-supervised ML techniques are a subset of unsupervised learning methods which can learn basic visual features from large scale unlabeled data without using any human-annotated labels [30]. Although this idea was initially conceived by Jürgen in 1989 [31] in the context of solving a credit assignment problem [32], it has recently been applied to various computer vision problems.

Currently, there are multiple sophisticated DL architectures and large-scale datasets which make convolutional networks the state-of-art algorithm for many computer-vision tasks. However, the collection of large-scale datasets is very expensive and time-consuming. In order to get a sense of the size of a "large-scale" dataset, ImageNet [33] one of the most widely used imaging datasets for pre-training DL networks contains about 1.3 million labeled images covering 1,000 classes and each image in a class is manually labeled by a human. Collecting a large-scale video dataset is even more expensive due to the added temporal dimension, and annotating these large-scale datasets is extremely labor intensive. One popular example is the Kinetics dataset [34] which consists of 500,000 videos of 600 classes of human action with each lasting about 10 seconds (Figure 5.2).

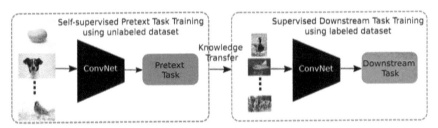

FIGURE 5.2   A general framework for self-supervised learning.

Source: Inspired by Ref. [30].

Note: A convolutional neural network is trained to learn the visual features and solve a pre-text task. These learned features then serve as a pre-trained model and together with the labeled dataset, the model is fine tuned to solve a downstream computer vision task.

To avoid manually performing these annotations, it was proposed to use self-supervised ML methods that can learn the visual features from unannotated large-scale datasets. One approach is to train a convolutional neural network to solve a pre-defined 'pretext task' using pseudo-labels (i.e., labels computed directly from the input raw data). The learned parameters of the convolutional neural network then serve as the pre-trained model and can be fine-tuned for downstream tasks. Figure 5.2 shows the framework for performing self-supervised learning. These methods differ from supervised learning in terms of labels. While supervised learning methods require a data pair X_i and y_i where y_i is human annotated labels, self-supervised methods also require a data pair X_i and p_i where p_i are pseudo labels generated for a pre-defined pretext task and are not labeled by humans.

An example of a pretext task is to predict the relationship between different patches of images, when each patch is extracted from an image. Doersch, Gupta, and Efros [35] formulates this task as predicting the relative position of an image patch with respect to the other patches as shown in Figure 5.3. Thus, the model needs to learn the spatial context of the features to define the relationship. An example in the context of medical image processing is discussed in Section 5.4.2.

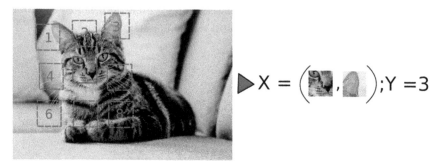

FIGURE 5.3   An example of pretext task, where the goal is to learn the patch representations. *Source:* Inspired by Ref. [35].

Note: The algorithm receives two patches, an anchor patch (represented in blue) and an input query patch (any one of the red patches). The goal of the algorithm is to classify the correct relative position of query patch with respect to the anchor patch from the eight possible spatial locations (represented in red). In the above example, if the algorithm is given the anchor patch corresponding to nose and eye (represented in blue) and query patch as an ear, then the algorithm is expected to classify that the ear corresponds to patch 3.

5.2.3   DEEP LEARNING (DL)

Classical ML models are generally unable to process the data in its raw form. Instead, they require domain expertise to extract relevant features from the raw data. In DL these features are instead learned directly from the raw data as part of the model. DL methods transform the raw data into an internal representation which the learning system can use to make predictions [17].

When applied to medical image analysis, one of the most important classes of DL models are convolutional neural networks (CNN), also referred to as ConvNet. Similar to the perceptron, CNNs are inspired by biological processes, as the connectivity pattern between the connected computational units, or neurons, resembles the actual neuronal organization of animal visual cortex. A CNN is made up of a number of layers and each layer is made up of neurons that may have learnable parameters (also known as weights) and biases. Figure 5.4 shows different kind of layers used to build a CNN. A CNN has very few connections between the layers and tends to preserve the spatial relationship in the data. The input of a CNN is arranged as a grid structure which is fed through different kinds of layers [36]. The computations on these layers are non-linear and transform the raw data into different levels of representation. For example, low level features learn the edges in an image, while high level representations learn abstract features in an image. The name "convolutional neural network" indicates that the network uses convolutions rather than matrix multiplication to learn these representations [37]. Figure 5.4 shows an example of a convolution operator on a 2D tensor. ConvNet consists of a sequence of layers and every layer transforms the output of previous layers using differentiable functions, also known as activation functions. Examples of ConvNet architectures and activation functions are described in Tables 5.1 and 5.2, respectively.

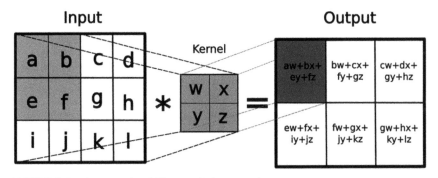

FIGURE 5.4   An example of 2D convolution operation.

Source: Inspired by Ref. [37].

Note: Imagine a 4 × 3 input image and a 2 × 2 filter (also referred to as kernel) which convolves the input volume by shifting one unit at a time. The amount by which the filter shifts is called its stride. In the convolution operation, the sum of element wise multiplication of the filter (represented by purple color) and the underlying image (represented by orange) is stored as the corresponding convolution output at a particular location (represented by green).

TABLE 5.1 High-Level Descriptions of Popular Convolutional Neural Networks

Name	Description
AlexNet [14]	This architecture is named after the first author Alex Krizhevsky. It has 7 layers and won the ImageNet large scale visual recognition competition (ILSVRC) 2012. The algorithm is able to classify an image into one of the 1,000 classes (e.g., dogs, cats, etc.), with high accuracy
VGG [38]	This architecture popularized the concept of using smaller filter kernels. This is achieved by increasing the number of layers to 19 thus training deep net-works.
GoogLeNet [39]	This architecture introduced a creative concept of stacking the layers in CNN. It consists of multiple inception modules [39] with multiple filters of different sizes. These are applied to the input and the results are concatenated. Rather than using fully connected layers at the end, this architecture uses global average pooling, thus reducing the number of parameters. This network won the ILSVRC 2015.
ResNet [15]	ResNet is the short name for residual network. This CNN architecture has skip connection, i.e., there are shortcut connections from one layer to other non-adjacent layers. The skip connections preserve information and helps in training deeper networks. The 152-layers ResNet won the ILSVRC 2015.
DenseNet [40]	This architecture is built on the idea of ResNet but with an exception that the layers are concatenated rather than connected. This architecture is suited for smaller datasets.
U-Net [16]	This architecture is also based on ResNet. It takes the input image and down-samples it using traditional CNN before up-sampling to the original size. During this process, it concatenates features from the down-sampling and up-sampling paths. This is a popular network for image segmentation applications.
GAN [41]	A generative adversarial network (GAN) consists of two neural networks: a generative and a discriminative network. The generative network creates samples and the discriminative network classifies the samples as they come from the generative network or training data.

As illustrated in Figure 5.5, a CNN largely consists of three main types of layers:

1. **Convolutional Layers:** These layers contain a set of parameterized filters (generally of dimension 3 × 3) that are convolved with the input image to extract features. These filters are referred to as kernel or feature detectors. Since these filters share the same weights across the entire input domain, the number of parameters

TABLE 5.2 Popular Activation Functions Used in the ConvNets

Name	Plot	Equation
Sigmoid		$f(x) = \dfrac{1}{1+e^{-x}}$
Tanh		$f(x) = \dfrac{e^{x} - e^{-x}}{e^{x} + e^{-x}}$
ReLU		$f(x) = max(0, x)$
ELU		$f(x) = \begin{cases} \alpha(e^{x} - 1) & for\ x <= 0 \\ x, & x > 0 \end{cases}$
SoftMax		$f(x) = \dfrac{e^{x}}{\sum e^{x}}$

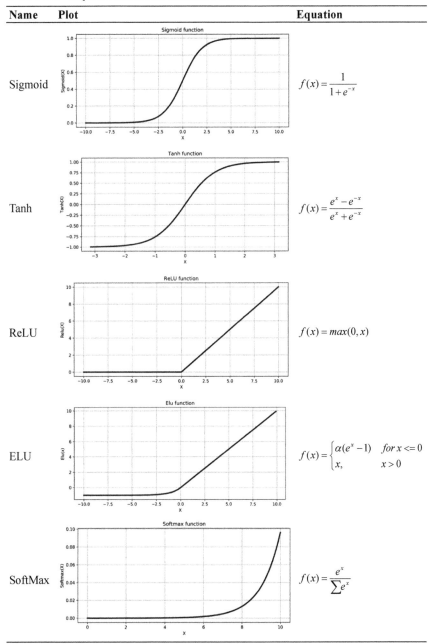

Abbreviations: Rectilinear unit (ReLU); exponential linear unit (ELU).

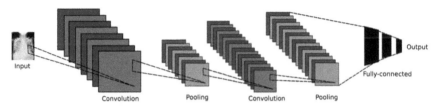

FIGURE 5.5   An example architecture of a convolutional neural network.
Source: Inspired by Ref. [42].

is significantly reduced. The motivation behind weight sharing is that if a filter is trained to detect a feature (e.g., vertical lines), then it can be used to detect the same feature anywhere in the input image and there is no need to re-train the filter. Once the filter is convolved with the entire input image it is fed to a non-linear activation function to produces a map known as feature map, or activation map. Mathematically, if a convolutional layer accepts an input volume with size (W_1, H_1, D_1), the spatial extent of the filter is F, stride S and padding P (padding refers to adding layers of zeros around the border of the input volume), then it produces an output volume with size (W_2, H_2, D_2) where $W_2 = \dfrac{W_1 - F + 2*P}{S} + 1$, $H2 = \dfrac{H_1 - F + 2*P}{S} + 1$ and $D_2 = K$ as shown in Figure 5.4.

2. **Pooling Layers:** These layers reduce the spatial dimensions of the input volume by replacing the output of a layer with summary statistics of nearby outputs. This layer makes the representation invariant to any small translations in the input data, making the network robust to small variations. More generally, a pooling layer accepts an input of volume size (W_1, H_1, D_1) and produces an output volume of size (W_2, H_2, D_2) where $W_2 = \dfrac{W_1 - F}{S} + 1$, $H_2 = \dfrac{H_1 - F}{S} + 1$ and $D_2 = D_1$. There are typically four types of pooling layers: (i) maximum pool, which computes the maximum within a rectangular neighborhood; (ii) average pool, which computes the average of a rectangular neighborhood; (iii) weighted pool, which computes the weighted average based on the distance from the central pixel; and (iv) L2-norm, which computes the L2-norm of a rectangular neighborhood.

3. **Fully Connected Layers:** In these layers all the nodes (or neurons) in the previous layer are connected to every neuron in the next layer. These layers are typically placed at the end of the network and compute the final output. Fully connected layers have activation functions (e.g., SoftMax functions) which uses the high-level features generated from convolutional and pooling layers to make the final prediction.

These DL architectures can be easily implemented in software frameworks like Keras [43], Tensorflow [44]; and Pytorch [45]. Each of these frameworks are open source and currently under active development. They can run on both central processing unit (CPU) and GPUs, but the latter make computations faster.

5.3 DATA FORMAT AND SOFTWARE

In this section we discuss common data formats used to store medical imaging data, as well as a variety of freely available software packages and resources that can be used to analyze them. Medical image file formats, like regular image file formats (e.g., .png and .jpeg), provide a standardized way to store image information. However, in contrast to many other types of images, medical image data often consist of more than two dimensions. For example, computed tomography (CT) imaging data consists of series of multi-slice two dimensional images that together make up a 3D volume, and functional magnetic resonance imaging (fMRI) data consists a dynamic series of volumes that together make a four-dimensional object (x, y, and z directions in space and time).

The data files are typically stored as binary data, using either 8 or more-bit integers (see Table 5.3). The data files also include information about the data, or meta-data, such as data dimensions, spatial resolution, pixel depth, and datatype. For example, a magnetic resonance image data file includes information related to the pulse sequence used (i.e., timing information, flip angle, number of acquisitions) A positron emission tomography (PET) image data file contains information about the radio-pharmaceutical agent injected and the body weight of the patient. A summary of standard file formats is shown in Table 5.3.

Most MRI scanners save the reconstructed image data as a DICOM (digital imaging and communications in medicine) file with the extension.'dcm.'

TABLE 5.3 Summary of Various File Formats Used in Medical Imaging [47]

Format	Header	Extension	Data Types	Year
Analyze	Fixed-length: 348-byte binary format	.img .hdr	Unsigned integer (8-bit), signed integer (16-, 32-bit), float (32-, 64-bit), complex (64-bit)	1986
DICOM [46]	Variable length binary format	.dcm	Signed and unsigned integer, (8-, 16-bit; 32-bit only allowed for radiotherapy dose), float not supported	1992
MINC	Extensible binary format	.mnc	Signed and unsigned integer (from 8- to 32-bit), float (32-, 64-bit), complex (32-, 64-bit)	1992
NIFTI	Fixed-length: 352-byte binary format	.nii	Signed and unsigned integer (from 8- to 64-bit), float (from 32- to 128-bit), complex (from 64- to 256-bit)	2001

Abbreviations: DICOM: Digital imaging and communications in medicine; MINC: medical imaging network common data format; NIFTI: neuroimaging informatics technology initiative.

The imaging data is stored slice-wise and contains large amounts of meta-data, including patient information and information about image acquisition settings. Usually, there are multiple DICOM files for each data acquisition and therefore, it is necessary to convert to other file format like NIFTI (neuroimaging informatics technology initiative) for data analysis. Brain imaging data structure (BIDS) [48] is a standard for organizing and describing the collected MRI dataset. It helps to organize DICOM images in a standard directory structure and describe the data collected during the experiment, making it easier to share the data and associated analysis pipelines. Recently, BIDS format has been extended to organize magneto-encephalography (MEG) [49], intracranial electroencephalography (iEEG) [50] and electroencephalography (EEG) [51] dataset.

In recent years there has been multiple initiatives to openly share medical imaging data between researchers. The goal of these initiatives is to maximize the contribution of research projects, improve research practice by enabling others to use the data to build new analysis methods and answer new questions, enhance reproducibility, and reduce the cost of conducting science [52]. In addition, these initiatives acquire data from different sites, scanners, imaging protocols making them a rich resource of diverse dataset which is helpful for improving science. Examples of freely available neuroimaging datasets include autism brain imaging data exchange (ABIDE) [53], functional connectomes project (FCP) [54], human connectome project (HCP) [55], OpenfMRI [56], and Adolescent Brain Cognitive Development (ABCD) [57]. These datasets include structural MRI, and task-based and resting state fMRI data. Some also include diffusion tensor imaging (DTI) data along with each subject's phenotype data. Other medical image datasets include the ChestX-ray8 dataset [58] which contains over 100,000 X-ray images and the UK Biobank (http://www.ukbiobankeyeconsortium.org.uk) which apart from neuroimaging data, also provides open access to retinal imaging. Prior to sharing these data, they are de-identified and no protected health information is shared.

Finally, there are multiple free open-source software packages used for medical image processing. The most popular package for neuroimaging (including fMRI, PET, SPECT, EEG, and MEG processing) is statistical parametric mapping (SPM) [59], a MATLAB based toolbox that allows for end-to-end analysis. AFNI (analysis of functional neuroimages) [60] is another comprehensive package for analysis of anatomical and function MRI. It has many built-in functions written in C, Python, R, and shell scripts. Its installation files are zipped in a binary package which can be easily installed in any operating system (OS). FSL (FMRIB software library) [61] is another package that can be used to analyze fMRI, MRI, and DTI data. It is also written in C and consists of a series of programs for pre-processing, conducting statistical analysis, and visualizing the results. FreeSurfer [62] is another C based software package that studies the surface of brain, especially the cortical and sub-cortical anatomy using structural, functional MRI, DTI, and PET. 3D Slicer [63] is an open-source platform for medical image analysis (e.g., image registration and segmentation) across multiple modalities including MRI, CT, Ultrasound, nuclear medicine, and microscopy. Furthermore, there is a python-based pipeline namely Nipype (Neuroimaging in Python Pipelines and Inter-faces) [64], which provides an interface to the above packages within a single framework. Thus, a user in Nipy (https://nipy.org/) can interactively explore algorithms from different packages and combine them to process data faster.

5.4   APPLICATIONS

5.4.1   IMAGE RECONSTRUCTION

The main objective of image reconstruction is to create high-quality medical images for clinical purpose at minimum cost and risk to the patient. The discovery of the X-ray fundamentally changed the practice of medicine and allowed physicians for the first time to study the anatomy of living patients without making incisions. When X-rays pass through the body a fraction of them are absorbed or scattered (based on the density of the structure), while the rest reach a photographic plate placed on the opposite side of the body. Depending on the fraction of X-rays that reach the plate, the intensity of the photographic plate changes. As dense structures lead to decreased intensity, which darkens the photographic plate, bones appear clearly in the resulting images.

   A fundamental problem with X-ray is that they are 2D projection of a 3D object. This implies that if a surgical tool is left on top of the patient's body, then the X-ray image would fail to identify if the tool is inside or outside the body [65]. This problem led to the development of advanced imaging modalities which could provide a cross-sectional 3D image of the body. For example, CT provided improved anatomical information, and PET provided additional metabolic information about the body. Later, imaging modalities like MRI provided increasingly detailed anatomical images, while functional MRI (fMRI) provided information about brain activity across the brain in response to a stimulus. Together, CT, PET, and fMRI have provided researchers with valuable insights into human brain function. In this section, we will review DL method used for reconstructing MRI image.

During MRI acquisition, the sensor encodes an intermediate representation of an object in the sensor domain [67]. A basic MRI acquisition setup and the role of the image reconstruction process is shown in Figure 5.6. The hardware includes a gradient magnetic field and radio-frequency (RF) coil which collectively serve as an encoding device. During MRI acquisition, the tissue being scanned is placed in a strong magnetic field (around 3 Tesla). This aligns the protons that are normally randomly oriented within the water nuclei of the tissue. This alignment (also referred to as magnetization) is perturbed through the application of external RF pulses. The subsequent loss of magnetization produces signal, and the average signal across the issue can be measured by the RF coil. When this signal is digitized using an analog to digital converter (ADC), it is referred to as raw data or k-space data (or data in the sensor domain). By altering the gradient, it is possible to create a spatially varying phase across the tissue, allowing measurement of weighted averages of the signal from the tissue where the weight depends on the applied gradient manipulation.

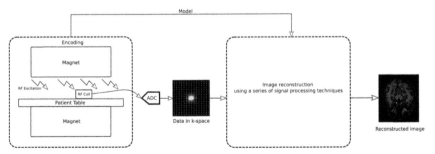

FIGURE 5.6   The basic MRI reconstruction process.

Source: Inspired by Ref. [66].

Note: The MRI hardware with RF coil collectively encodes the imaging data in k-space. The role of reconstruction process is to use multiple signal processing steps to transform the k-space data into a reconstructed image.

The goal of image reconstruction is to process the k-space data to reconstruct an image of the object of interest. The reconstruction process relies on an encoding model, which keeps track of the gradient manipulations used to create the k-space measurements. However, due to imperfections in the measurement system, the model may deviate from the actual encoding process, resulting in artifacts in the reconstructed image [66]. To circumvent these issues, Zhu et al. [67] proposed a data driven DL framework, automated transform by manifold approximation (AUTOMAP), which learns the reconstruction mapping between the input data in sensor-domain to the final reconstructed data in the image-domain. During the training process, the network implicitly learns a low-dimensional joint manifold of the data in both domains, as shown in Figure 5.7, thus capturing a robust representation of the input data.

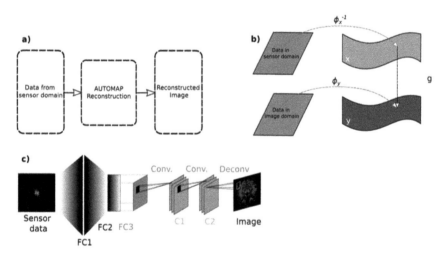

FIGURE 5.7   An example deep learning architecture for image reconstruction.
Source: Inspired by Ref. [67].
Note: (a) The basic image reconstruction framework using AUTOMAP. (b) A low dimensional joint manifold $x \times y$ is learned in a supervised manner, mapping the data in sensor domain to the data in image domain. (c) The network architecture of AUTOMAP uses fully connected layers (FC1–FC3) with hyperbolic tangent activation functions and convolutional layers (C1 and C2) with ReLU activation functions.
Abbreviations: Conv.: convolution operations; Deconv.: deconvolution operations.

Mathematically, AUTOMAP manifold learning can be explained as a two-fold process. Let $\tilde{x}$ be the noisy observation of x, the data in sensor domain. The first step is to learn the stochastic projection operator onto $\mathcal{X}: p(\tilde{x}) = P(x \mid \tilde{x})$ which removes the noise. Once x is obtained, the next task is to reconstruct $f(x)$ by computing a reconstruction mapping function which minimizes the reconstruction error. Considering an ideal situation in which the data in sensor domain is noiseless, the input data for $i - th$ observation can be denoted as $\{y_i, x_i\}_{i=1}^n$, where x_i represents a $n \times n$ set of input parameters and y_i indicates the $n \times n$ underlying images. An additional assumption is that there exists a smooth and homeomorphic function such that $y = f(x)$ and $\{x_i\}_{i=1}^n$, $\{y_i\}_{i=1}^n$ lie on unknown smooth manifolds $\mathcal{X}$ and $\mathcal{Y}$, respectively [67]. Thus, collectively, a joint manifold can be defined as $\mathcal{M}_{\mathcal{X},\mathcal{Y}} = \mathcal{X} \times \mathcal{Y}$ within which the entire dataset $(x_i, y_i)_{i=1}^n$ lies and can be formulated as:

$$\mathcal{M}_{\mathcal{X},\mathcal{Y}} = \{(x, f(x)) \in \mathbb{R}^{n^2} \times \mathbb{R}^{n^2} \mid x \in \mathcal{X}, f(x) \in \mathcal{Y}\} \tag{3}$$

Note, the $(x, f(x))$ is described using a regular Euclidean coordinate system. Equivalently, it can be described using an intrinsic coordinate system of $\mathcal{M}_{\mathcal{X},\mathcal{Y}}$ as $(k, g(k))$ such that there is a homeomorphic mapping ϕ (ϕ_x, ϕ_y) = between $(x, f(x))$ and $(k, g(k))$. This can be represented as $x = \phi_x$ (k) and $f(x) = \phi_y \circ g(k)$. Therefore, in the manifold learning, the network learns the diffeomorphism g between $\mathcal{X}$ and $\mathcal{Y}$. Consequently, $f(x)$ can be written as:

$$f(x) = \phi_y \circ g \circ phi_x^{-1}(x) \tag{4}$$

Together, AUTOMAP determines both a between-manifold projection g from $\mathcal{X}$ to $\mathcal{Y}$ (the manifold of output images), and a manifold mapping ϕ_y to project the image from manifold $\mathcal{Y}$ back to the representation in Euclidean space [67]. AUTOMAP provides a novel image reconstruction framework which learns a reconstruction function that best represents the data in a low-dimensional feature space using manifold learning and sparse convolutional filters. This helps to solve a general reconstruction problem and can be applicable to a broad range of imaging modalities.

5.4.2 IMAGE REGISTRATION

Image registration [68] is a process for aligning different images into a common reference system. It allows information to be integrated from different images of the same scene taken at different times, or from different viewpoints or sensors. In medical image processing image registration is one the most important step, as it helps combine data from various imaging modalities like MRI, PET, and SPECT (single-photon emission computed tomography) to extract more detailed information about the patient. Mathematically, image registration seeks to align two images (a sensed image to a reference image) by finding a transformation function ϕ that minimizes the dissimilarity between the images [69]. Mathematically, we can express this as follows:

$$\phi = argmin[dissim(I_R, I_S)] \tag{5}$$

where; I_R and I_F denote the reference and sensed image, respectively; and *dissim* is any function which compares two images (e.g., intensity sum of squared distance (SSD), mean square distance (MSD), correlation ratio, normalized cross-correlation or mutual information).

There are many tool kits that have been developed for image registration, including SimpleITK, 3D Slicer and ANTS. The methods implemented in these tool kits work by optimizing the similarity between two images and updating the transformation metrics accordingly. However, these methods tend to be very slow, requiring many iterations to optimize the cost function $dissim(I_R, I_S)$ from Eqn. (5).

As an alternative, a self-supervised based learning method can be applied to solve this problem. It uses a large amount of unlabeled data to create 'pseudo labels' and train an affine image registration network (AIRNet) [70] which directly estimates the transformation matrices. This approach has many advantages over conventional methods. First, it is two orders of magnitude faster. Second, it achieves better overall performance. The method is flexible and can incorporate multiple imaging modalities. Additionally, the network is able to identify discriminative features in data from an unseen imaging modality, not even part of the training dataset (Figure 5.8).

The AIRNet network has two main components, an encoder and regression part. The encoder part has two pathways, one for each input image. Each pathway has 2D and 3D filters which help learn the discriminative features of the images. The filters on each pathway share weights which reduces the number of parameters in the model. The pathways from both images are concatenated and fed into the regression part, which helps find the non-linear relationship between the discriminative features and transformation matrix.

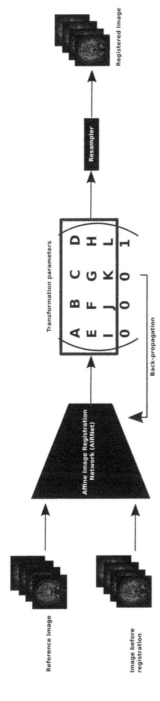

FIGURE 5.8 Image registration using affine image (AIRNet) [70]. The network takes a reference image and the image to be registered as input and minimizes the loss function (mean squared error of the estimated transformation matrices). The weights of the network are tuned using backpropagation. The fully trained network directly outputs the transformation matrices used for image registration. The letters in the transformation matrix represent the amount of translation, scaling, rotation, and shearing.

Source: Inspired by Ref. [70].

During training the loss between the actual and estimated transformation parameters is minimized. However, there is no ground truth for the actual transformation matrix. Thus, a self-supervised learning algorithm can be used to train the network. Specifically, for every MRI a random transformation matrix is generated. For example, one could allow parameters such as angle of rotation about the z-axis to vary between -0.8 to 0.8 radians, translation can vary from -0.15 to 0.15 along the y-axis and -0.2 to 0.2 along z-axis. Similarly, constraints can be added for scaling and shearing. The random transformation matrix can be multiplied with the image to produce an affine transformed version of the image. Using this technique, one can easily generate unlimited synthetic data and pseudo labels that can be used to train the network. The affine transformed image can now be used as an input along with the original image defined as I_i and I_{ref} respectively. The random transformation matrix serves as the ground truth label. Thus, the loss function is defined as [70]:

$$Loss = \frac{1}{K}\sum_{i=1}^{K}\left|\left|\overline{j^{[i]}} - AIRNet\left(I_{ref^{[i]}}, I_{i^{[i]}}; \vec{W}\right)\right|\right|_2^2 \qquad (6)$$

where; K is the total number of training samples; $\vec{W}$ represents the network weights tuned during the training phase; $I_{ref^{[i]}}$ and $I_{\overline{i^{[i]}}}$ are a training pair at the $i-th$ iteration; and $\overline{j^{[i]}}$ is their corresponding labels, and AIRNeT is the network itself.

5.4.3   SURGICAL ROBOTS

Surgical robots (also known as robot-assisted surgery or robotic surgery) allow doctors to perform surgery with more precision, control, and flexibility than using conventional techniques. Currently, the ™da Vinci surgical system is the state-of-the-art technique for robotic surgery. It includes specialized "arms" for holding the instruments, camera, magnified screen, computer console, and other surgical accessories. To operate using the robotic system, the doctor makes a tiny incision in the body and inserts the surgical instruments and camera. The doctor uses the external computer console to guide the instrument to the surgical site and perform surgery. The doctor is always in control of the robot and the system responds to directions provided by the doctor.

Currently, a robotic assisted surgery faces a number of challenges [71]:

1. **Pre-Operative Planning:** The robot can be used to assist the surgeon in determining whether surgery is required. In addition to multi-model imaging techniques like MRI and PET, the surgeon receives information about the patient such as gender, age, genetic history, and symptoms. The potential inclusion of less informative diagnostic data can make it challenging for the surgeon to make important diagnostic decisions. Here, a robotic intervention can help the surgeon predict the success of the surgery based on existing information about the patient.

2. **Intra-Operative Guidance:** Computer assisted intra-operative guidance is required to better visualize and localize the surgical site. Often, a surgeon needs robotic assistance to understand the 3D shape of the site. Intra-operative 3D reconstruction allows for the acquisition of 3D images. However, this process can be both slow and produce low-resolution images. Incorrect or limited information during a critical surgery could be life threatening. Also, given the dynamic nature of the tissue, improved real time localization of the surgical site will help the surgeon perform a successful surgery.

3. **Surgical Robots:** This as an autonomous system must be capable of perceiving the complex in-vivo environment, make decisions to perform the task with safety, precision, and highest accuracy. Additionally, the system must have precise control over the maneuvering of the surgical instruments. The system must be intelligent enough to interact with human and learn from human demonstration of complicated surgery.

The introduction of AI to the fields of healthcare and surgery has already made strong contributions to imaging and navigation for pre-operative planning and intra-operative guidance. One recent example is SuPer deep [72], a surgical perception framework, which provides geometrical information about the entire surgical scene, including the surgical tool and deforming agent (e.g., the tissue).

The framework uses two different DL models, one for localizing the surgical tool in the image frame and performing feature extraction, and another for tracking tissue manipulation. In order to localize the surgical tool, the framework uses DeepLabCut [73, 74], which is a deep convolutional network (deep-learning) that combines two key ingredients from algorithms for object recognition and semantic segmentation, namely pre-trained residual networks (ResNets) and deconvolutional layers. The framework uses stereo images and trains the model on pre-defined features for each view. For each frame, DeepLabCut localizes the surgical tool and predicts the 2D coordinates of features. The 2D detections are combined to construct a 3D view of the scene (Figure 5.9).

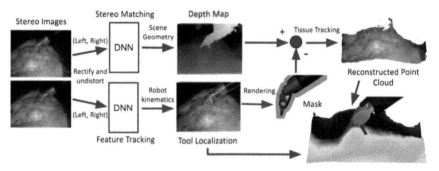

FIGURE 5.9 A block diagram of the SuPer deep surgical framework. The framework integrates two deep learning pipelines, each taking a pair of stereo images as input. One pipeline is used to extract the 2D features of the surgical tool and estimate its location in 3D space, while the other is used to estimate the disparity map.

Source: Figure appears with permission from Ref. [72].

Simultaneously, a DL network referred to as pyramid stereo matching network (PSMNET) [75] is used to match the stereo images and estimate a disparity map of each pixel. This results in a dense depth image as shown in Figure 5.9. Briefly, PSMNET is a DL network which extends pixel-level features to region-level features with different scales of receptive fields. These combined global and local features help estimate a reliable disparity map using the stereo images.

Despite these well recognized benefits, current robotic platforms and frameworks have certain limitations. The SuPer deep surgical framework [72] is computationally expensive and running two DL networks in parallel makes the entire framework very slow. Additionally, the framework is limited to very specific tissue types and surgical tools. Moreover, setting up a robotic surgical unit for any institution is extremely expensive, ranging in cost from $1 million to $2.5 million [76], and regular maintenance can cost around $138,000 per year [77]. Most importantly, robotic systems lack a sensory feedback system which allows the system to send a signal to the surgeon in case it interacts with any rigid structure (e.g., tool-to-tool collisions). The surgeon needs to guess the presence of the surgical tool and this can lead to improper needle handling or suture breaks.

However, with the increasing use of robotics, AI is set to transform the future of surgery. The development of more sophisticated sensorimotor functions with different levels of autonomy can provide the system with the ability to adapt to constantly changing and patient-specific in vivo environments.

5.4.4   IMAGE SEGMENTATION

Image segmentation is the process of partitioning an image into multiple parts. For example, one may be interested in segmenting an MRI into gray matter, white matter, and cerebral spinal fluid. A good segmentation should have the following properties [78]:

- The segmented regions should be homogeneous and uniform with respect to its texture and color;
- Segmented regions should be connected and must not have any holes;
- Adjacent regions of the segmented objects must have significantly different attributes;
- Boundaries of segmented object should not be ragged and must be spatially accurate.

Achieving all of these properties is extremely challenging, especially in the context of medical images. This is due to the non-rigid nature of the object and its deformations, as well as the large-scale changes between body parts of same patient (e.g., renal artery) and statistical variations in normal and disease ground truth data. Nevertheless, a good segmentation plays an important task in many analyzes and pre-operative planning. For example, in the case of renal artery segmentation using CT image, if the segmentation can provide a mask of the renal artery, it gives the surgeon an opportunity to locate the exact position to add clamps [79].

In order to achieve a good renal artery segmentation, He et al. [79] proposed a semi-supervised framework which is robust to inter-anatomy variations. The framework uses a deep prior anatomy strategy (DPA) which uses an unsupervised algorithm to extract features and embed feature representations in a supervised model to adapt the segmentation to the anatomy. Thus, the combination of unsupervised training in stage 1 and supervised training in stage 2 makes this framework a semi-supervised learning strategy as discussed in Section 5.2.2.1.

Figure 5.10 shows the proposed framework "DPA-DenseBiasNet" [79] for renal artery segmentation. It consists of three main structures which work together in two stages.

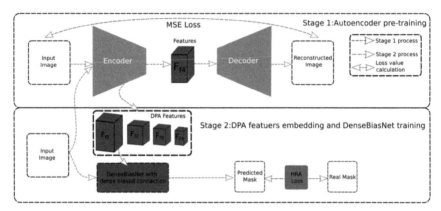

FIGURE 5.10   A general overview of DPA-DenseBiasNet for image segmentation.

Source: Inspired by Ref. [79].

Note: In the first stage, an autoencoder is trained on unlabeled images where the encoder unit extracts the DPA features which help to learns the representation of anatomical features. In second stage, the DenseBiasNet is trained using the DPA features which adapts to the anatomical variations in the image by optimizing the HRA loss function.

1. **Auto-Encoder:** It is an unsupervised neural network used for representational learning [80]. It tries to learn a function such that the output $\hat{x}$ is similar to the input x. By adding additional constraints on the network, an Autoencoder can compress the data and discover interesting structure. It has two main units, an encoder unit which learns the function and reduces the dimensions of the input data and a decoder unit which uses the encoded representations to reconstruct the input data. In stage 1, the Autoencoder is trained using a large number of unlabeled data to learn the representations of numerous anatomical features. This strategy is robust, suitable for thin structures and can be generalized to other anatomical structures.

2. **DenseBiasNet:** It is a variant of the standard 3D U-Net [81] which takes dense biased connections as the connectivity pattern. It compresses and transmits the feature maps in each layer to every

forward layer [79]. The main advantage of this network is the dense connectivity patterns. These patterns help with information flow and gradient flow which are completely transmitted in the network. Additionally, the network is robust to differences in scale. Together, this makes training faster. In stage 2, DenseBiasNet receives the representations from the encoder network of the Autoencoder. The final renal artery segmentation is achieved by training the DenseBiasNet on labeled data and optimizing the loss function.

3. **Loss Function:** A cross-entropy loss function is used for segmentation, which can mathematically be defined for binary segmentation as – $(y log(p) + (1 - y) log(1 - p))$. Here y is a binary indicator (0 or 1) and p is the predicted probability for belonging to each class. However, if the loss function is applied to the entire image, it will cause a huge class imbalance problem, as there are many regions which can be easily segmented but others that are harder to segment. This problem can lead to poor overall performance. Thus, He et al. [79] define a novel sampling strategy which uses segmentation quality to sample regions that are more difficult to segment while ignoring the rest. This helps improve the overall segmentation performance. This strategy is important in segmenting thin structures (like interlobular arteries in the renal artery segmentation) which are relatively hard to identify but play a vital role during renal surgery. The authors define this strategy as hard region adaptation (HRA) loss function.

This framework provides excellent results in renal artery segmentation and greatly improves the efficiency of pre-operative planning for renal surgeries. The low-level features are more sensitive to thin structures, so once the features are transmitted to the next layers, the framework achieves excellent results in these structures which are traditionally difficult to segment. This framework serves as an example of using semi-supervised learning strategy in medical image segmentation. Although, it requires higher processing time, renal segmentation does not currently require real time application.

5.5   FUTURE DEVELOPMENTS

Just as human and animal learning is largely unsupervised (i.e., we learn the structure of world by observing it), similarly we expect unsupervised learning to play an increasingly important role in medical image processing [17]. In particular, we expect that major progress will come through developing systems that combine representation learning with complex reasoning.

Though much progress has been made combining AI and medical image processing, there remains a disconnect between how humans think and the way computers make predictions, i.e., a black box problem [82]. There is a difference in opinion among the experts about the severity of this issue. Some believe that if an algorithm can consistently improve a doctor's performance, there is no need to understand the mechanism behind the algorithm. Others point to the fact that subtle confounding factors can alter an algorithm's performance and impact its generalizability.

For example, [83] found that the performance of their ML algorithm achieved greater than 90% accuracy in identifying pneumonia in lung X-ray data collected at their institute. However, it performed relatively poorly in data collected from other institutes. This seems to imply that there is some confounding factor related to the institutes data that does not carry-over to data from other institutes. Uncovering these factors can be difficult when the mechanism behind the algorithm is unclear. This raise concerns that the data used to create an algorithm can introduce biases that the developers may not have initially considered. This is of course particularly problematic in situations where a person's life is at risk. Additionally, there are a number of legal concerns related to the black box problem. If a ML algorithm fails to make a correct diagnostic decision, and the doctor makes an incorrect decision, it is difficult to determine whether the doctor or the algorithm is at fault.

Ultimately, we expect that the availability of large open-source data-sets and open-source ML libraries will help solve many of these problems. One of the main advantages of open-source data is that it provides data from many different sites, populations, and data acquisition protocols giving the researcher an opportunity to answer a research question in an unbiased way. Nevertheless, a careful evaluation of the data needs to be done prior to sharing, as there is a potential risk of re-identification of the subject [84]. This also opens the door to the concept of federated ML [85]. This is a class of distributed systems where copies of ML algorithms are distributed to the nodes where the data reside. This way the data remains with the owner, while still enabling the training of the algorithm [86]. We expect that federated learning approaches will become widely used as a next generation privacy preservation technique in industry and medical applications [87].

We expect AI to continue to play an important role in medical image processing, as combined with a doctor's experience it can help improve outcomes. However, we doubt AI will replace doctors anytime soon, as human biology is complex. Instead, we strongly believe that with the development of new ML tools and techniques, the training of doctors will definitely change. Thus, we agree with the experts that state that AI will not replace the doctors, but doctors who use AI will replace the doctors who do not use AI [82].

KEYWORDS

- artificial intelligence
- artificial neural network
- central processing unit
- convolutional neural networks
- exponential linear unit
- generative adversarial network

REFERENCES

1. Nilsson, N. J., (2014). *Principles of Artificial Intelligence*. Morgan Kaufmann.
2. Carbonell, J. G., Michalski, R. S., & Mitchell, T. M., (1983). An overview of machine learning. In: *Machine Learning* (pp. 3–23). Elsevier.
3. Nilsson, N. J., (1996). *Introduction to Machine Learning: An Early Draft of a Proposed Textbook*. USA; Stanford University.
4. Cohen, W. W., et al., (1996). Learning rules that classify e-mail. In: *AAAI Spring Symposium on Machine Learning in Information Access* (Vol. 18, p. 25).
5. Asadi, N., & Sadrodini, M., (2010). *Employing Data Mining to Identify Cancer Risk Factors and Determine the Optimal Treatment in Namazi Hospital Cancer Database* (Vol. 16, pp. 17–18).
6. Turing, A. M., (1950). Computing machinery and intelligence. *Mind, 50*, 433–460.
7. McCulloch, W. S., & Pitts, W., (1943). A logical calculus of the ideas immanent in nervous activity. *Bull. Math. Biophys., 5*(4), 115–133.

8. Hebb, D. O., (1949). *The Organization of Behavior: A Neuropsychological Theory*. J. Wiley; Chapman & Hall.

9. Rosenblatt, F., (1962). *Principles of Neurodynamics*. Washington, D. C Spartan.

10. Kononenko, I., (2001). Machine learning for medical diagnosis: History, state of the art and perspective. *Artif. Intell. Med., 23*(1), 89–109.

11. Quinlan, J. R., (1986). Induction of decision trees. *Mach. Learn., 1*(1), 81–106.

12. Muggleton, S., & De Raedt, L., (1994). Inductive logic programming: Theory and methods. *J. Log. Program., 19*, 629–679.

13. Good, I. J., (1950). *Probability and the Weighing of Evidence*. Edited by Good, I. J., et al.

14. Krizhevsky, A., Sutskever, I., & Hinton, G. E., (2012). ImageNet classification with deep convolutional neural networks. In: *Advances in Neural Information Processing Systems* (pp. 1097–1105).

15. He, K., Zhang, X., Ren, S., & Sun, J., (2016). Deep residual learning for image recognition. In: *Proceedings of the IEEE Conference on Computer Vision and Pattern Recognition* (pp. 770–778).

16. Ronneberger, O., Fischer, P., & Brox, T., (2015). U-net: Convolutional networks for biomedical image segmentation. In: *International Conference on Medical Image Computing and Computer-Assisted Intervention* (pp. 234–241).

17. LeCun, Y., Bengio, Y., & Hinton, G., (2015). Deep learning. *Nature, 521*(7553), 436–444.

18. Kumar, A., Bi, L., Kim, J., & Feng, D. D., (2020). Machine learning in medical imaging. In: *Biomedical Information Technology* (pp. 167–196). Elsevier.

19. Cortes, C., & Vapnik, V., (1995). Support-vector networks. *Mach. Learn., 20*(3), 273–297.

20. Tolman, E. C., (1932). *Purposive Behavior in Animals and Men*. Univ of California Press.

21. Craik, K. J. W., (1952). *The Nature of Explanation, 445*. CUP Archive.

22. Dayan, P., Sahani, M., & Deback, G., (1999). Unsupervised learning. *MIT Encycl. Cogn. Sci.*, 857–859.

23. Barlow, H. B., (1989). Unsupervised learning. *Neural Comput., 1*(3), 295–311.

24. Chapelle, O., Scholkopf, B., & Zien, A., (2009). Semi-supervised learning [book reviews]. *IEEE Trans. Neural Netw., 20*(3), 542–552.

25. Zhu, X., & Goldberg, A. B., (2009). Introduction to semi-supervised learning. *Synth. Lect. Artif. Intell. Mach. Learn., 3*(1), 1–130.

26. Merz, C. J., Clair, D. S., & Bond, W. E., (1992). Semi-supervised adaptive resonance theory (smart2). In: *[Proceedings 1992] IJCNN International Joint Conference on Neural Networks* (Vol. 3, pp. 851–856).

27. Hosmer, Jr. D. W., (1973). A comparison of iterative maximum likelihood estimates of the parameters of a mixture of two normal distributions under three different types of sample. *Biometrics*, 761–770.

28. Dempster, A. P., Laird, N. M., & Rubin, D. B., (1977). Maximum likelihood from incomplete data via the EM algorithm. *J. R. Stat. Soc. Ser. B Methodol., 39*(1), 1–22.

29. Belkin, M., Niyogi, P., & Sindhwani, V., (2006). Manifold regularization: A geometric framework for learning from labeled and unlabeled examples. *J. Mach. Learn. Res., 7*, 2399–2434.

30. Jing, L., & Tian, Y., (2020). Self-supervised visual feature learning with deep neural networks: A survey. *IEEE Trans. Pattern Anal. Mach. Intell.*

31. Schmidhuber, J., (1990). *Making the World Differentiable: On Using Self-Supervised Fully Recurrent Neural Networks for Dynamic Reinforcement Learning and Planning in Non-Stationary Environments.* pp. 1–26.

32. Minsky, M., (1961). Steps toward artificial intelligence. *Proc. IRE, 49*(1), 8–30.

33. Deng, J., Dong, W., Socher, R., Li, L. J., Li, K., & Fei-Fei, L., (2009). ImageNet: A large-scale hierarchical image database. In: 2009 *IEEE Conference on Computer Vision and Pattern Recognition* (pp. 248–255).

34. Kay, W., et al., (2017). *The Kinetics Human Action Video Dataset.* ArXiv Prepr. ArXiv170506950.

35. Doersch, C., Gupta, A., & Efros, A. A., (2015). Unsupervised visual representation learning by context prediction. In: *Proceedings of the IEEE International Conference on Computer Vision* (pp. 1422–1430).

36. Lundervold, A. S., & Lundervold, A., (2019). An overview of deep learning in medical imaging focusing on MRI. *Z. Für Med. Phys., 29*(2), 102–127.

37. Goodfellow, I., Bengio, Y., & Courville, A., (2016). *Deep Learning.* MIT press.

38. Simonyan, K., & Zisserman, A., (2014). *Very Deep Convolutional Networks for Large-Scale Image Recognition.* ArXiv Prepr. ArXiv14091556.

39. Szegedy, C., et al., (2015). Going deeper with convolutions. In: *Proceedings of the IEEE Conference on Computer Vision and Pattern Recognition* (pp. 1–9).

40. Huang, G., Liu, Z., Van, D. M. L., & Weinberger, K. Q., (2017). Densely connected convolutional networks. In: *Proceedings of the IEEE Conference on Computer Vision and Pattern Recognition* (pp. 4700–4708).

41. Goodfellow, I., et al., (2014). Generative adversarial nets. In: *Advances in Neural Information Processing Systems* (pp. 2672–2680).

42. LeCun, Y., Bottou, L., Bengio, Y., & Haffner, P., (1998). Gradient-based learning applied to document recognition. *Proc. IEEE, 86*(11), 2278–2324.

43. Chollet, F., et al., (2015). *Keras.*

44. Abadi, M., et al., (2016). TensorFlow: A system for large-scale machine learning. In: *12th {USENIX} Symposium on Operating Systems Design and Implementation ({OSDI} 16)* (pp. 265–283).

45. Paszke, A., et al., (2017). *Automatic Differentiation in PyTorch.* pp. 1–4.

46. Bidgood, Jr. W., & Horii, S. C., (1992). Introduction to the ACR-NEMA DICOM standard. *Radiographics, 12*(2), 345–355.

47. Larobina, M., & Murino, L., (2014). Medical image file formats. *J. Digit. Imaging, 27*(2), 200–206.

48. Gorgolewski, K. J., et al., (2016). The brain imaging data structure, a format for organizing and describing outputs of neuroimaging experiments. *Sci. Data, 3*(1), 1–9.

49. Niso, G., et al., (2018). MEG-BIDS, the brain imaging data structure extended to magnetoencephalography. *Sci. Data, 5*(1), 1–5.

50. Holdgraf, C., et al., (2019). IEEG-BIDS, extending the brain imaging data structure specification to human intracranial electrophysiology. *Sci. Data, 6*(1), 1–6.

51. Pernet, C. R., et al., (2019). EEG-BIDS, an extension to the brain imaging data structure for electroencephalography. *Sci. Data, 6*(1), 1–5.

52. Poldrack, R. A., & Gorgolewski, K. J., (2014). Making big data open: Data sharing in neuroimaging. *Nat. Neurosci., 17*(11), 1510.

53. Di Martino, A., et al., (2014). The autism brain imaging data exchange: Towards a large-scale evaluation of the intrinsic brain architecture in autism. *Mol. Psychiatry, 19*(6), 659–667.

54. Finn, E. S., et al., (2015). Functional connectome fingerprinting: Identifying individuals using patterns of brain connectivity. *Nat. Neurosci., 18*(11), 1664.

55. Van, E. D. C., et al., (2013). The WU-minn human connectome project: An overview. *Neuroimage, 80*, 62–79.

56. Poldrack, R. A., et al., (2013). Toward open sharing of task-based fMRI data: The OpenfMRI project. *Front. Neuroinformatics, 7*, 12.

57. Casey, B., et al., (2018). The adolescent brain cognitive development (ABCD) study: Imaging acquisition across 21 sites. *Dev. Cogn. Neurosci., 32*, 43–54.

58. Wang, X., Peng, Y., Lu, L., Lu, Z., Bagheri, M., & Summers, R. M., (2017). Chestx-ray8: Hospital-scale chest x-ray database and benchmarks on weakly-supervised classification and localization of common thorax diseases. In: *Proceedings of the IEEE Conference on Computer Vision and Pattern Recognition* (pp. 2097–2106).

59. Penny, W. D., Friston, K. J., Ashburner, J. T., Kiebel, S. J., & Nichols, T. E., (2011). *Statistical Parametric Mapping: The Analysis of Functional Brain Images*. Elsevier.

60. Cox, R. W., (1996). AFNI: Software for analysis and visualization of functional magnetic resonance neuroimages. *Comput. Biomed. Res., 29*(3), 162–173.

61. Jenkinson, M., Beckmann, C. F., Behrens, T. E., Woolrich, M. W., & Smith, S. M., (2012). FSL. *Neuroimage, 62*(2), 782–790.

62. Fischl, B., (2012). FreeSurfer. *Neuroimage, 62*(2), 774–781.

63. Fedorov, A., et al., (2012). 3D slicer as an image computing platform for the quantitative imaging network. *Magn. Reson. Imaging, 30*(9), 1323–1341.

64. Gorgolewski, K., et al., (2011). Nipype: A flexible, lightweight and extensible neuroimaging data processing framework in python. *Front. Neuroinformatics, 5*, 13.

65. Lindquist, M. A., (2016). From CT to fMRI: Larry shepp's impact on medical imaging. *Annu. Rev. Stat. Its Appl., 3*, 1–19.

66. Hansen, M. S., & Kellman, P., (2015). Image reconstruction: An overview for clinicians. *J. Magn. Reson. Imaging, 41*(3), 573–585.

67. Zhu, B., Liu, J. Z., Cauley, S. F., Rosen, B. R., & Rosen, M. S., (2018). Image reconstruction by domain-transform manifold learning. *Nature, 555*(7697), 487–492.

68. Zitova, B., & Flusser, J., (2003). Image registration methods: A survey. *Image Vis. Comput., 21*(11), 977–1000.

69. Cao, X., Fan, J., Dong, P., Ahmad, S., Yap, P. T., & Shen, D., (2020). Image registration using machine and deep learning. In: *Handbook of Medical Image Computing and Computer Assisted Intervention* (pp. 319–342). Elsevier.

70. Chee, E., & Wu, Z., (2018). *AirNet: Self-Supervised Affine Registration for 3D Medical Images Using Neural Networks*. ArXiv Prepr. ArXiv181002583.

71. Zhou, X. Y., Guo, Y., Shen, M., & Yang, G. Z., (2019). *Artificial Intelligence in Surgery*. ArXiv Prepr. ArXiv200100627.

72. Lu, J., Jayakumari, A., Richter, F., Li, Y., & Yip, M. C., (2020). *SuPer Deep: A Surgical Perception Framework for Robotic Tissue Manipulation using Deep Learning for Feature Extraction*. ArXiv Prepr. ArXiv200303472.

73. Mathis, A., et al., (2018). DeepLabCut: Marker less pose estimation of user-defined body parts with deep learning. *Nat. Neurosci., 21*(9), 1281.

74. Nath, T., Mathis, A., Chen, A. C., Patel, A., Bethge, M., & Mathis, M. W., (2019). Using DeepLabCut for 3D marker less pose estimation across species and behaviors. *Nat. Protoc., 14*(7), 2152–2176.

75. Chang, J. R., & Chen, Y. S., (2018). Pyramid stereo matching network. In: *Proceedings of the IEEE Conference on Computer Vision and Pattern Recognition* (pp. 5410–5418).

76. Barbash, G. I., & Glied, S. A., (2010). New technology and health care costs—the case of robot-assisted surgery. *N. Engl. J. Med., 363*(8), 701–704.

77. Hussain, A., Malik, A., Halim, M., & Ali, A., (2014). The use of robotics in surgery: A review. *Int. J. Clin. Pract., 68*(11), 1376–1382.

78. Haralick, R. M., & Shapiro, L. G., (1985). Image segmentation techniques. *Comput. Vis. Graph. Image Process., 29*(1), 100–132.

79. He, Y., et al., (2020). Dense biased networks with deep priori anatomy and hard region adaptation: Semi-supervised learning for fine renal artery segmentation. *Med. Image Anal.*, 101722.

80. Hinton, G. E., & Salakhutdinov, R. R., (2006). Reducing the dimensionality of data with neural networks. *Science, 313*(5786), 504–507.

81. Çiçek, Ö., Abdulkadir, A., Lienkamp, S. S., Brox, T., & Ronneberger, O., (2016). 3D U-Net: Learning dense volumetric segmentation from sparse annotation. In: *International Conference on Medical Image Computing and Computer-Assisted Intervention* (pp. 424–432).

82. Reardon, S., (2019). Rise of robot radiologists. *Nature, 576*(7787), S54.

83. Zech, J. R., Badgeley, M. A., Liu, M., Costa, A. B., Titano, J. J., & Oermann, E. K., (2018). Variable generalization performance of a deep learning model to detect pneumonia in chest radiographs: A cross-sectional study. *PLoS Med., 15*(11), e1002683.

84. Schwarz, C. G., et al., (2019). Identification of anonymous MRI research participants with face-recognition software. *N. Engl. J. Med., 381*(17), 1684–1686.

85. Konečnỳ, J., McMahan, B., & Ramage, D., (2015). *Federated Optimization: Distributed Optimization Beyond the Datacenter.* ArXiv Prepr. ArXiv151103575.

86. Kaissis, G. A., Makowski, M. R., Rückert, D., & Braren, R. F., (2020). Secure, privacy-preserving and federated machine learning in medical imaging. *Nat. Mach. Intell.*, 1–7.

87. Rieke, N., et al., (2020). *The Future of Digital Health with Federated Learning.* ArXiv Prepr. ArXiv200308119.

CHAPTER 6

CONVENTIONAL AND ADVANCED MAGNETIC RESONANCE IMAGING METHODS

RUPSA BHATTACHARJEE and SNEKHA THAKRAN

Center for Biomedical Engineering, Indian Institute of Technology Delhi, New Delhi, India
E-mail: snekhathakran@gmail.com (S. Thakran)

ABSTRACT

Magnetic resonance imaging (MRI), also called nuclear MRI, is a special acquisition method that creates minute images of the body. The scan uses a magnetic field (weak/medium or strong: 0.2, 1.5, 2, 7T) and radio waves to create images of parts of the body. These parts are otherwise not visible well with X-rays, CT scans, or ultrasound. One such example would be: it can help clinicians to see joints inside the body, ligaments, cartilage, muscles, etc. This is useful for detecting various sports injuries. MRI helps to diagnose multiple disorders and can examine the internal body, such as spinal cord injuries, strokes, aneurysms, multiple sclerosis, tumors, and eye or inner ear problems. It is also widely used in research to measure brain structure and function, in addition to multiple other things. An MRI is an extremely accurate method of disease understanding. After the other testing methods fail to provide sufficient enough information to confirm a patient's diagnosis, MRI is generally the last resort. In the head, bleeding/swelling can indicate brain trauma. Several other abnormalities are also detected via MRI, and some examples would be: aneurysms, stroke, tumors of the brain, tumors, and inflammation of the spine. MRI scans help neurosurgeons to visualize the brain anatomy, spinal cord integrity in trauma situations, diseases associated with the spine, such as vertebrae or intervertebral disc issues. In heart and aorta structures, in detecting aneurysms or tears, MRI proves to be a valuable component. Especially in glands and organs of the abdomen, MRI can provide valuable information. In the case of joints structures, soft tissues, and bones of the body, a high-resolution MRI can provide minute details that are otherwise not visible in any other modality. MRI scans help to confirm the surgical outputs, progress tracking, treatment planning, and several such cases. In this chapter, the authors summarized conventional and advance Magnetic Resonance Imaging Methods.

6.1   MAGNETIC RESONANCE IMAGING (MRI)

MRI [1] is an imaging technique used in medical science to visualize the detailed internal structure of an object under examination. Felix and Edward in 1946 [23] first discovered the principle of nuclear magnetic resonance (NMR), on which MRI is based on. NMR was in general, is used for physical analysis of molecules. Around 1971, Damadian et al. experimented and observed NMR relaxation times are seen to be varied in healthy vs. tumor tissues. This came as a motivation for researchers to utilize NMR for disease detection. Paul C. Lauterbur first demonstrated MRI in 1972. Richard Ernst came up with the concepts of using frequency encoding, phase encoding, and Fourier Transform in MRI in 1975. These are still the building blocks of state-of-the-art MRI systems presently available. The area of MRI is expanding with the advancements in technology. In the present world, MRI has become one of the go-to choices for many such disease detection situations. MRI is typically beneficial compared to other imaging modalities (CT/X-Ray) in the fact that it is non-invasive, involves non-ionizing radiation. On the other hand, it can provide excellent and high-resolution soft tissue contrast that still has no competition.

6.1.1   BASIC PRINCIPLE OF MRI

As we all know, atomic nuclei generally consist of protons and neutrons. These are found with a net positive charge. They possess another important property, called "spin" that depends on how many protons the nucleus has. Spin can be imagined as the nucleus spinning around its axis. This example is an analogy. In real the nucleus does not spin around in the classical meaning, rather it constitutes of parts. These parts can induce a certain magnetic moment and can generate a local magnetic field (think of north and south poles) [2].

Here we considered a dipolar magnet. The quantum mechanics of the dipolar magnet is almost synonymous to the spinning of objects in classical mechanics. Imagine the dipole to be a bar magnet, with north/south poles aligned along the rotational axis (Figure 6.1).

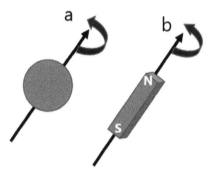

FIGURE 6.1 Example of the spin of a nucleus. (a) It behaves like a bar magnet inducing a magnetic field; (b) north and south are represented by N and S, respectively. The direction of the magnetic field is indicated by the arrows.

When we apply an externally strong magnetic field (B0), the nucleus aligns with the field. This alignment can either be parallel or anti-parallel to the external field. Now, if we take a liquid solution (comprising of multiple nuclear spins), place the solution in B0 field; the spins present will be in either of two conditions: a) low-energy state (this is aligned parallel to B0 field) or b) high-energy state (orientated this is aligned anti-parallel to B0 field). In examples where we consider solids or liquids, there will be a tendency present within most of the spins to align parallel to the B0 field.

However, in considering a bar magnet, it will completely align either parallel or antiparallel to the B0 field. The nucleus is rotating and will have an angular momentum as well. Thereby it will rotate or precess. This precession happens around the B0 axis. We can often visualize this particular behavior as a wobbling motion of a gyroscope when the gyroscope is placed under the Earth's magnetic field. Such visualization can explain the utility of "spin" in a quantum mechanical phenomenon. The precession velocity around the B0 field direction is also known as the Larmor frequency.

This is expressed via the Larmor equation:

$$\omega_0 = \gamma B_0 \tag{1}$$

Spinning nuclei, present within the external magnetic field have the potential to be excited. It requires the application of a second magnetic field (B1). B1 is generated by radiofrequency pulse (RF) perpendicular to B0. Each RF pulse is short and lasts about a few microseconds only. The spinning nuclei absorb the RF energy transmitted via these RF pulses. Energy absorption and relaxation can cause a transition from higher to lower energy levels and vice versa.

This particular energy that is being absorbed or emitted by spinning nuclei causes to induce a typical voltage change. If we suitably tune a coil of wire, we can pick up the voltage change. Subsequently, it can be amplified and then presented as "free-induction decay" (FID).

When the RF pulses are not being transmitted anymore, the relaxation process happens, leading the system into thermal equilibrium. Each of the nuclei present will resonate at a unique frequency which is the characteristic feature of the nuclei depending on its position in the human body, also affected by surrounding nuclei. Even if the nuclei are in the same B0 field, the resonance frequency will be different.

The energy level difference between two typical nuclear states of the spinning nucleus the same as the transitional energy requirement between levels. It also depends on the weakness/strength on which the nuclei are placed, B0.

$$\Delta E = \frac{\gamma h B_0}{2\pi} \tag{2}$$

When we apply the RF pulse at the particular resonant frequency of the spin, it generates an FID at the end of the process [2].

In reality, multiple FIDs are needed to obtain and multiple RF pulses are used to achieve so. Average of the multiple FID signals used to obtain a higher signal-to-noise ratio (SNR). Basically, this FID (averaged signal) is a time-domain signal. This time-domain signal consists of contributions made from various nuclei within that same strong magnetic B0 field environment. For example, there can be free water molecules and $_1$H which are bound to specific tissues present in the human body. Each of them will contribute differently to the time domain FID.

Here Fourier transformation (another well-known mathematical technique) is implied which converts the time domain signal into the frequency domain, separating out the various frequency components mixed within the time domain FID. The final output can be an MRI image, it can also be a spectrum of frequencies having biochemical information (depending upon the process and requirement) as demonstrated in Figure 6.2.

RF pulse at resonant frequency can be applied to generate an FID [2]. Multiple RF pulses are applied in practice which produce multiple FIDs. Averaging multiple FIDs results in improved SNR. Different nuclei within the environment like free water and 1 H bound to tissue contribute towards signal-averaged FID which is a time-domain signal. The resolution of signal-averaged FID can be done by using Fourier transformation (mathematical process) into an image (MRI) or a frequency spectrum, providing biochemical information (Figure 6.2).

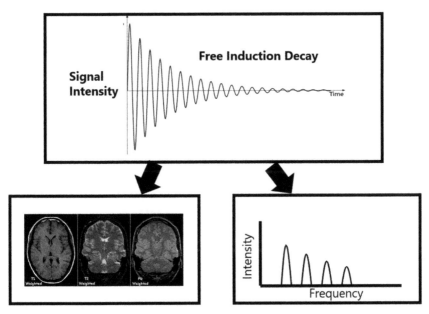

FIGURE 6.2 Use of free induction decay (FID), and Fourier transformation in generating MR images or MR spectroscopy signal.

6.2 CONVENTIONAL MRI TECHNIQUES

To use different contrast mechanisms (T1 and T2) different combinations of sequences and pulse designs are implied. Two types of time constants are defined as T1 and T2 are spin-lattice-relaxation-constant, and spin-spin-relaxation-constant respectively. By conventional MRI we mean the different types of sequences that are routinely being used in an MRI scanner on a daily basis. Examples include proton density (PD), fluid-attenuated-inversion-recovery (FLAIR), and T2-weighted-spin-echo sequence. These are useful in the brain specifically for detecting various types of abnormality. For lesion activity ad growth gadolinium contrast media is injected and T1-weighted images are acquired [2]:

1. **T1 Weighted Image:** In this kind of image the contrast dependence factor is mainly on the differences in the T1 relaxation time values amongst various tissues such as fat and/or water. Here the TR has the control of recovery that each vector can achieve before subsequent RF pulse is applied for excitation. Thereby, if we look at achieving T1 weighting, we need to keep the TR short. The value should be such that by that time no fat/water tissues have good enough time to fully return to their previous magnetization state before excitation. If we keep the TR is long enough that fat and water both can recover back to B0, they recover longitudinal magnetization completely. If this occurs, the T1 relaxation mechanism will be complete in fat and water in both the tissues. The differences in T1 relaxation times of fat and water would not be visible on the image, thus it will be of no use.

2. **T2 Weighted Image:** In this the contrast mostly is depended on the differential T2 times amongst various tissues such as water and fat. How much T2 decay is allowed to happen (before we receive the signal) depends on the TE. If we want to achieve T2 weighting, we must allow TE to be long enough. It allows both water and fat tissues to decay. In case the TE time is kept short, neither water nor fat will be able to fully decay. The T2 time differential between them would not be highlighted.

3. **Proton Density Image:** In the case of this, the main factor that determines the contrast of the image is the differential number of protons/volumes under acquisition. In every MR image PD weighting is to some extent, present. To fully achieve PD weighted image contrast, T1 and T2 effects have to be eliminated in such a way that only PD contrast is visible. The way to do so is to keep a long TR, allowing water and fat tissues to be fully recovered from their longitudinal T1 magnetization effect, i.e., no T1 weighting.

Besides, a short TE needs to be put to ensure fat and water both do not decay and there is no T2 weighting present as well.

4. **Fluid Attenuated Inversion Recovery (IR):** It is commonly used sequence. Flair sequence and T2-weighted image have a lot of similarities except the TE and TR times which are very long for Flair. Because of this the normal CSF fluid is attenuated and made dark while the abnormalities stay bright. The sequence has high sensitivity towards pathology and makes it easier to differentiate between CSF and an abnormality (Figure 6.3).

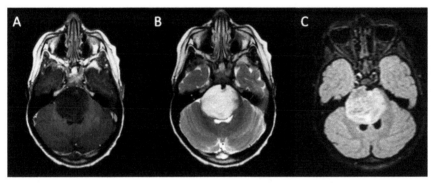

FIGURE 6.3   Example of T1-weighted, T2-weighted and flair images [19] in brain.

Source: Image by Katherine E. Warren (2012). https://creativecommons.org/licenses/by/3.0/us/

6.3   ADVANCED MRI TECHNIQUES

A number of advanced MR imaging methods have been adopted in clinical practices in addition to the conventional ones and others are subject of intense research [2]. The conventional MR imaging sequences provide anatomic information but the advanced techniques provide a lot more information like generating physiologic data and giving information on chemical composition. Some advanced methods include: diffusion-weighted-imaging (DWI), perfusion imaging, diffusion-tensor-imaging (DTI), blood oxygen level-dependent (BOLD) imaging, MR spectroscopy, chemical exchange saturation transfer (CEST), magnetization transfer constant (MTC), and the susceptibility weighted imaging (SWI):

1. **Perfusion Imaging:** Blood passing within the vascular network of the body is explained through perfusion imaging. There are various

techniques developed to quantify brain perfusion [14, 16] through measuring hemodynamic parameters such as cerebral blood volume, cerebral blood flow, etc. [2]. This measurement is done by serially measuring the concentration of a tracer in the area of interest. Some examples of such exogenous tracers would be iodinated radiographic contrast material, iced saline solution, and radionuclides. Now with the developments of MR imaging, paramagnetic contrasts are being used as exogenous contrast. Magnetically labeled blood can also be used to act as endogenous tracer contrast. Previously mentioned hemodynamic parameters are measured serially from a certain tissue using a model. This kind of model is based on knowledge of how the contrast is infused. There are two ways, bolus injection or contrast infusion. It also requires assumptions on how the pharmacokinetic properties of the contrast behave in that specific tissue. Such assumptions would include various parameters such as the volume of distribution, diffusibility (from the intravascular to extravascular space), and equilibrium half-life of the agent. Three major types of MR perfusion imaging involve arterial spin labeling, T2*-weighted dynamic susceptibility contrast, and T1-weighted dynamic contrast-enhanced methods. MR Perfusion guides clinicians with quantitative biomarkers help to characterize tumor vascularity, helps in tumor grading. It also has the potentials to predict prognosis and can quantify therapeutic efficacy as well. This is very crucial as we are now looking at anticancer agents which can specifically target some aspects of lesion biology. These findings are generally not found by assessing general conventional MR imaging.

2. **Diffusion-Weighted-Imaging (DWI):** It is aimed to detect the random movements of water protons inside biological tissues [12]. Water molecules diffuse relatively differently in the different kinds of tissue environments like water molecules in the brain cerebrospinal fluid diffuse freely in an isotropic fashion compared to intracellular and extracellular space where water molecules feel restriction and hindrance in their path caused by the different tissue structures and boundaries. Spontaneous movements, also known as Diffusion, rapidly becomes limited in ischemic brain tissue. Throughout Ischemia, the sodium-potassium pump shuts down which results in intracellular sodium accumulation. The osmotic gradient results in shift of water from extracellular to intracellular

space. The signal is very bright on DWI due to restricted intracellular water movement. Because of this, DWI is a very sensitive method for detecting acute stroke.

3. **Diffusion Tensor Imaging (DTI):** It is the simplest method/mathematical model to fit the diffusion-weighted MRI signal [18]. DTI model fitting is based on the assumption that diffusing water molecules in the biological tissues follow a gaussian distribution. DTI model fitting provides information about the tissue microstructures at the millimeter level in the form of measured degree of anisotropy of diffusion and diffusivities. The degree of anisotropy is called fractional anisotropy (FA) and diffusivity along the largest dimension of the ellipsoid tensor is axial diffusivity (AD), perpendicular to the largest dimension of the ellipsoid tensor is radial diffusivity (RD) and mean diffusivity (MD). DTI has been widely used in the clinical routines to examine the tissue structures. DTI also provides ways to track the nerve fibers in the brain based on the measured FA, which is called tractography. The concept that water molecules follow a gaussian distribution in complex biological tissues is an oversimplification of the actual scenario which can be tackled by the diffusion-kurtosis-imaging (DKI). DKI is the extension of the DTI model and encounters the non-gaussian nature of the diffusing molecules.

4. **MR Spectroscopy:** It essentially represents a quantitation of brain chemistry [7]. The nuclei that are most commonly being used, are ^1H (proton), ^{31}P (phosphorus) and ^{23}Na (sodium). Proton spectroscopy is comparatively easiest to perform. It also provides much higher SNR than any other nuclei like sodium or phosphorus. Proton MRS Studies are completed within 10–15 minutes on an average. Therefore, it can easily be added on to routine MR imaging protocols in practice. MRS is generally used to observe biochemical changes in epilepsy, stroke, tumors, metabolic disorders, infections as well as neurodegenerative diseases serially. They need interpretation and should be correlated with the MRI images before reaching a final diagnosis. The MR Image is constructed by taking the total signal from all the protons in a voxel. All the signals cannot be used for MRS as Water and Fat have very high peaks which can make other metabolite peaks invisible due to scaling. Since Fat and Water are not areas of focus, they can be safely eliminated. Fat is removed

by placing the MRS voxel within the brain, away from the fat in scalp and bone marrow. Water suppression can be done by either IR (inversion recovery) or CHESS (chemical-shift selective) technique. The suppression techniques used along with STEAM or PRESS pulse sequence acquisition. Data is then used with Fourier transform to separate the signal into individual frequencies. Protons resonate at different frequencies inside different molecules due to the local electron cloud affecting the magnetic field.

5. **Blood Oxygen Level Dependent Imaging [17]:** How the blood flows in the brain are highly controlled. This control mechanism happens locally with response to oxygen and carbon dioxide (CO_2) tension given to cortical tissue. In response to a specific task, activity of a specific region of the cortex is increased. This is followed by local capillaries having extraction fraction of oxygen. This further leads to a drop in oxygenated hemoglobin (oxyHb) initially. Also, an increase in local CO_2 and deoxygenated hemoglobin(deoxy-Hb) is observed. After a certain time-lag, cerebral blood flow increases, in turn delivering an additional amount of oxygenated hemoglobin. The deoxyhemoglobin gets washed away. Imaging is done of the large rebound in local tissue oxygenation. There is a fundamental difference in paramagnetic properties of oxy-Hb and deoxy-Hb due to which fMRI can detect this change. Oxygenated hemoglobin is non-paramagnetic but deoxygenated hemoglobin is paramagnetic so it causes local dephasing of protons thus reducing the signal intensity of the tissues in the immediate vicinity. There is an order of 1–5% change detected using heavily T2* weighted sequences.

6. **Chemical Exchange Saturation Transfer Imaging (CEST):** This imaging is another way of approaching MRI contrast relaxation [15]. Exogenous/endogenous compounds that either contain exchange-able protons or similar molecules. These compounds are selectively saturated, followed by indirect detection via water signal. This gives enhanced sensitivity.

In CEST MRI, phenomenon of magnetization transfer is observed not in semisolids, rather in mobile compounds. CEST and MTC both have contributions from chemical exchange and dipolar cross-relaxation. However, dipolar cross relaxation effects are neglected many a times in fast exchange situations. Unlike MTC, a significantly slow exchange (in terms of MR time scale) is needed in CEST. This is for protons under interest to have selective irradiation. Therefore, radio-frequency saturation is not the only factor for magnetic labeling. It can also be expanded with approaches which are slower frequency-selective. Some examples would be frequency labeling, inversion, or gradient dephasing.

7. **Magnetization Transfer Constant (MTC):** This MRI method [11] eliminates a part of the total signal in tissue, thus improving the contrast. Macromolecular protons are saturated by an Off-resonance radio frequency (RF) pulse. This makes them invisible due to ultra-short $T2^*$ relaxation times. Semi-solid tissues like brain parenchyma have reduced signal and more fluid components like blood have signals retained, e.g., When broad spectral lines are saturated, it may result in decrease in intensity of lines not directly saturated through the magnetization exchange between corresponding states. More closely coupled states will have greater intensity change. By using magnetization transfer techniques, gadolinium contrast-enhanced T1-weighted images and T2 weighted images provide better visibility for demyelinated brain and spine lesions. Off-resonance uses selection gradient during an off-resonance MTC pulse. On the arterial side of imaging volume, the gradient has a negative offset frequency. The net effect of this type of pulse is that the arterial blood outside the imaging volume will have more longitudinal magnetization, with the more vascular signal on entering the imaging volume. Off-resonance MTC saturates the venous blood but not the arterial blood. On resonance will saturate the bound water pool but has no effect on the free water pool and is the difference in T2 between the pools. Binomial pulses cause the magnetization of the free protons to unchanged.

The z-magnetizations reverts to original value. With a short T2, the spins of the bound pool experience decay destroying magneti-zation after the on-resonance pulse.

8. **Susceptibility-Weighted-Imaging (SWI):** It is a high-spatial-resolution 3D fully velocity corrected gradient-echo MRI sequence [4, 13]. Diamagnetic, Paramagnetic, and ferromagnetic compounds interact with the local magnetic field and distort it which alters the phase of local tissue resulting in loss of signal. Deoxyhemoglobin, ferritin, and hemosiderin are paramagnetic compounds whereas bone minerals and dystrophic calcifications are diamagnetic compounds.

Post processing step takes place after acquisition which comprises the application of a phase map to accentuate the directly observed signal loss. The images give typically the following: filtered phase, magnitude, SWI minimum intensity projection (mIP) and SWI (combined post-processed magnitude and phase) (Figure 6.4).

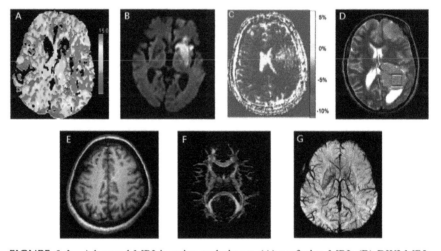

FIGURE 6.4   Advanced MRI imaging techniques. (A) perfusion MRI; (B) DWI-MRI; (C) CEST-MRI; (D) MRS-MRI; (E) FMRI; (F) DTI-MRI; and (G) SWI-MRI.

Source: A: Image by Shazia Mirza and Sankalp Gokhale. https://en.wikipedia.org/wiki/ Creative_Commons. B: Reprinted from Ref. [21]. https://creativecommons.org/licenses/ by/2.0/deed.en. C: Reprinted from Ref. [15]. https://creativecommons.org/licenses/ by/4.0/. D. Image by Dj manton at English Wikipedia At the University of Hull, Centre for Magnetic Resonance Investigations (http://www.hull.ac.uk/mri). https://creativecommons. org/licenses/by/3.0/deed.en. E. Reprinted from Ref. [20]. https://creativecommons.org/ licenses/by/2.5/deed.en. F: Image by Laurent Hermoye. G: Image by SBarnes. https:// creativecommons.org/licenses/by-sa/3.0/deed.en

6.4   CONCLUSION

Various researches are going on in the domain of MRI. Exploration of multiple contrasts and easier processing raises the expectation that down-the-line MRI will be available at a low cost, improving accessibility. MRCP and MRI of the abdomen, pancreas, liver, and brain, will become common for people to access. Works are also going on in the domain of increasing field strengths. 3T machines are routinely available now, 4T and 7T are not available as human scanners. But these higher strength machines are utilized for animal studies in general. Tissue contrast and higher resolution have the potential to increase the significance of MRI in diagnosis. Researchers are working on reducing the number of invasive diagnostic techniques. To conclude, it is very important to have a brief understanding of MRI in today's era, to know the physical; principles, and important imaging parameters. That will ensure the users and practitioners have complete in-depth knowledge of this state-of-the-art imaging modality. MRI is considered a powerful technique because it provides exquisite soft tissue and anatomic detail. Compared to other techniques like CT scans and x-rays, MRI has no risk of radiation exposure which is a major benefit.

KEYWORDS

- **diffusion-weighted-imaging**
- **fluid-attenuated-inversion-recovery**
- **free-induction decay**
- **magnetic resonance imaging**
- **nuclear magnetic resonance**
- **proton density**

REFERENCES

1. Elston, C. W., & Ellis, O. I., (1991). Pathological prognostic factors in breast cancer. I. The value of histological grade in breast cancer: Experience from a large study with long-term follow-up. *Histopathology, 19*, 403–410.
2. McRobbie, D., Moore, E., Graves, M., & Prince, M., (2006). Front matter. In: *MRI from Picture to Proton* (pp. I–VI). Cambridge: Cambridge University Press.
3. http://mriquestions.com/what-is-susceptibility.html (accessed on 30 September 2021).
4. Haacke, E. M., Mittal, S., Wu, Z., et al., (2009). Susceptibility-weighted imaging: Technical aspects and clinical applications, part 1. *AJNR Am. J. Neuroradiol., 30*, 19–30. PMID: 19039041.
5. Bloch, F., Hansen, W. W., & Packard, M. E., (1946). Nuclear induction. *Phys Rev., 69*, 127.
6. Westbrook, C., Roth, C. K., & Talbot, J., (2011). *MRI in Practice* (4th edn.). London: John Wiley & Sons, Inc.
7. Di Costanzo, A., Trojsi, F., Tosetti, M., et al., (2003). High-field proton MRS of human brain. *Eur. J. Radiol., 48*(2), 146–153.

8. Soher, B. J., Dale, B. M., & Merkle, E. M., (2007). A review of MR physics: 3 T versus 1.5 T. *Magn. Reson. Imaging Clin. N. Am., 15*(3), 277–290.

9. Rovira, A., Cordoba, J., Sanpedro, F., Grive, E., Rovira-Gols, A., & Alonso, J., (2002). Normalization of T2 signal abnormalities in hemispheric white matter with liver transplant. *Neurology, 59*(3), 335–341.

10. Rovira, A., Grive, E., Pedraza, S., Rovira, A., & Alonso, J., (2001). Magnetization transfer ratio values and proton MR spectroscopy of normal appearing cerebral white matter in patients with liver cirrhosis. *Am. J. Neuroradiol., 22*(6), 1137–1142.

11. Wolff, S. D., & Balaban, R. S., (1989). Magnetization transfer contrast (MTC) and tissue water proton relaxation in vivo. *Magn. Reson. Med., 10*(1),135–144.

12. Moseley, M. E., Kucharczyk, J., Mintorovitch, J., et al., (1990). Diffusion-weighted MR imaging of acute stroke: Correlation with T2 weighted and magnetic susceptibility-enhanced MR imaging in cats. *AJNR Am. J. Neuroradiol., 11*(3), 423–429.

13. Bhattacharjee, R., Gupta, R. K., Patir, R., Vaishya, S., Ahlawat, S., & Singh, A., (2019). Quantitative vs. semi-quantitative assessment of intra-tumoral susceptibility signals in patients with different grades of glioma. *J. Magn. Reson. Imaging.* doi: 10.1002/jmri.26786.

14. Thakran, S., Gupta, P. K., Kabra, V., Saha, I., Jain, P., Gupta, R. K., & Singh, A., (2018). Characterization of breast lesion using T1-perfusion magnetic resonance imaging: Qualitative vs. quantitative analysis. *Diagn. Interv. Imaging.* pii: S2211-5684 (18)30142-6.

15. Paech, D., Zaiss, M., Meissner, J. E., Windschuh, J., Wiestler, B., Bachert, P., et al., (2014) Nuclear over Hauser enhancement mediated chemical exchange saturation transfer imaging at 7 tesla in glioblastoma patients. *PLOS One, 9*(8), e104181. https://doi.org/10.1371/journal.pone.0104181.

16. Bhandari, A., Bansal, A., Singh, A., & Sinha, N., (2017). Perfusion kinetics in human brain tumor with DCE-MRI derived model and CFD analysis. *Journal of Biomechanics, 59*, 80–89.

17. Chen, F., Li, S., & Sun, D., (2018). Methods of blood oxygen level-dependent magnetic resonance imaging analysis for evaluating renal oxygenation. *Kidney Blood Press Res., 43*, 378–388.

18. Andrew, L. A., Jee, E. L., Mariana, L., & Aaron, S. F., (2007). Diffusion tensor imaging of the brain. *Neurotherapeutics, 4*(3), 316–329.

19. Warren, K. E., (2012). Diffuse intrinsic pontine glioma: Poised for progress. *Frontiers in Oncology, 2*, 205.

20. Kim, J., Matthews, N. L., & Park, S., (2010). An event-related FMRI study of phonological verbal working memory in schizophrenia. *PLOS One, 5*, e12068.

21. Shen, J., Xia, X., Kang, W., et al., (2011). The use of MRI apparent diffusion coefficient (ADC) in monitoring the development of brain infarction. *BMC Med Imaging, 11*(2). https://doi.org/10.1186/1471-2342-11-2.

22. Shazia, M., & Sankalp, G., (2016). *Neuroimaging in Acute Stoke.* pp. 1–38.

23. Felix and Edward (1946). 20-02 | The history of MR imaging. MRI NMR Magnetic Resonance. Essentials, introduction, basic principles, facts, history. The primer of EMRF/TRTF. (1946). www.magnetic-resonance.org. https://www.magnetic-resonance.org/ch/20-02.html (accessed 3 November 2021)

CHAPTER 7

DETECTION AND CLASSIFICATION OF BRAIN TUMORS FROM MRI IMAGES BY DIFFERENT CLASSIFIERS

J. V. BIBAL BENIFA,[1] JIPSA PHILIP,[2] and CHANNA BASAVA CHOLA[1]

[1]*Indian Institute of Information Technology, Kottayam, Kerala, India,*
E-mail: benifa.john@gmail.com (J. V. B. Benifa)

[2]*Rajiv Gandhi Institute of Technology, Kottayam, Kerala, India*

ABSTRACT

Brain tumor is the phenomenal growth of abnormal tissues in the human brain. A malignant tumor is known as a cancerous cell, and it is the major cause of death among people across the globe. Nowadays, the detection and classification of brain tumors from Magnetic Resonance Images (MRI) is a very crucial task. This chapter addresses various computing methods such as Edge Detection (ED), feature extraction, and classification techniques for the detection and classification of brain tumor regions from MRI datasets. Initially, MRI images are collected, and then various pre-processing steps such as filtering and edge detection are applied. Then, different edge detection methods such as Sobel ED, Robert's ED, Prewitt ED, Canny ED, Laplacian ED, and Laplacian of Gaussian (LoG) with sigma 3 are applied for the MRI datasets. Subsequently, segmentation techniques are applied for detecting tumor regions from MRI, and essential features are extracted using Discrete Wavelet Transform (DWT) method. Otsu's thresholding and K-means clustering segmentation methods are used for the investigation. Further, the support vector machines (SVMs), Naïve Bayes (NB), K-nearest neighbors (KNN), Back Propagation Neural Network (BPNN), and feedforward neural networks (FFNN) classifiers are employed for the MRI classification purpose. The experiments are conducted using the brain tumor dataset in the MATLAB 2019a software environment. The experimental results are analyzed in multiple dimensions, and it shows that the SVM with Otsu's thresholding method exhibits better performance with 86.11% accuracy during the classification.

7.1   INTRODUCTION

Human body is made up of plenty of cells and when the cell growth becomes out of control, the additional clustering of cells would be changed into tumor. Hence, the phenomenal growth of abnormal tissues within the brain is classified as brain tumor [1]. Diagnosis of exact tumor region is a complex process because often tumors are overlapped with the dense brain tissues [2]. MRI is a well-known technique that uses magnetic fields and radio waves to get detailed information about the tissues in a human body [3]. Brain tumors are broadly classified into two types, namely Benign and Malignant tumor as displayed in Figure 7.1. Though Benign and Malignant tumors are called cancerous cells, malignant tumors grow rapidly than the Benign counterpart and it is a life threatening one [4]. Hence, classification of the type of tumor is very important for the effective treatment of cancer patients. Moreover, brain tumor detection and treatment are a challenging task because of the complex structure, shape, and texture of tumor tissues. Therefore, advanced machine learning (ML) techniques should be incorporated to detect the tumor regions from MRI.

Every year, more than 1 million people are affected in each country with cancer related diseases. Among the index of cancerous death across the globe, the second leading cause of death is brain tumor [2]. Hence, effective diagnosis methods are required for the early detection of brain tumors. Segmentation is an effective method and it is a time-consuming process while it is performed manually. Therefore, computer aided automatic classification strategies need to be used for the accurate brain tumor detection and classification [5]. In this chapter, four classifiers are used for tumor classification and their performance is evaluated. For the classification process, initially brain MRI images are considered as input data and it is converted into gray scale images. Then, these images are subjected to various preprocessing steps in the MATLAB environment. Since the images contain a desirable amount of noise, suitable filtering techniques are applied.

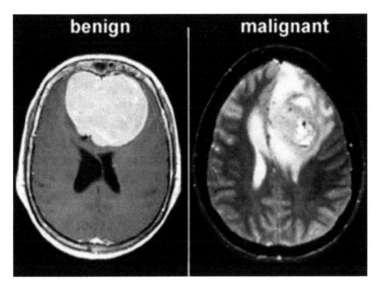

FIGURE 7.1   Benign and malignant tumor.

Median filtering is used as a reliable filtering technique to remove the noise from MRI dataset for the efficient output [6]. In the next preprocessing step, edge detection methods such as Prewitt [7]; Robert's [7]; Sobel [7]; Canny [8]; Laplacian [9]; LoG [9] are applied. Then, the segmentation process is initiated to segment the tumor regions from MRI images by Otsu's Thresholding [10] and K means clustering [11]. Totally, 13 features are extracted through DWT [12] and feature vectors are reduced by principal component analysis (PCA) [13]. Subsequently, classification is performed to classify the images into normal and abnormal tissues using the four classifiers namely, SVM [14], NB [15], KNN [16], and NN (BPNN and FFNN) [2, 17] classifiers. The remainder of the chapter has been divided into following sections: Section 7.2 summarizes the related works; Section 7.3 narrates the proposed methodology; Section 7.4 enumerates the results and discussion; and Section 7.5 contains the concluding remarks.

7.2   RELATED WORKS

The summary of brain tumor detection and classification works done by the researchers and the inferences from their studies are presented in this section. Telrandhe et al. [18] presented a brain tumor detection technique from MRI images by using K-Means segmentation and SVM [18]. Here, the collected images are subjected to several preprocessing techniques like median filtering and skull masking. Then, the Texture and color features are extracted and subsequently SVM is used for classification. Rashid and Mamun (2018) proposed a brain tumor detection method by anisotropic filtering for preprocessing. Then, SVM classifier categorizes the images into two normal and abnormal classes [19]. The morphological operations are applied finally to show the tumor regions exist in the original image.

Zulkoffli (2019) detected tumor regions from brain MRI by extraction of features using K-means clustering and morphological operations [20]. Publicly available online MRI dataset is utilized and the pre-processing stage includes the conversion of images into grayscale and applying filtering techniques. In the post-processing stage, K-means clustering algorithm and morphological operations are used to separate the tumor regions from MRI. Features like energy, contrast, kurtosis, correlation, and homogeneity along with the area and perimeter of the tumor are extracted for accurate calculation. Zaw et al. (2019) has collected the dataset from the Rembrandt database and it was applied with various preprocessing techniques like morphological operations and pixel subtraction. Further, the segmentation process is done based on the threshold to extract intensity and morphological features [21].

Kharrat et al. (2019) proposed a mathematical morphology-based method to improve the dissimilarities while detecting tumor from the brain MRI images. Wavelet decomposition and K-means algorithm are used for segmentation and classification of tumor images [22]. Özyurt et al. (2020) proposed a brain tumor detection method based on fuzzy C-means (FCM) clustering and convolution neural network (CNN) known as super resolution fuzzy-C-means (SR-FCM) CNN. This method segments the tumor regions using SR-FCM approach and the feature extraction is done using CNN architectures integrated with extreme learning machines (ELM) [23]. Kumar (2020) investigated a dolphin sine cosine algorithm (SCA) based on Deep CNN to improve the accuracy for brain tumor classification [24]. Here, the input MRI images are preprocessed and the segmentation process is done by a fuzzy deformable fusion model. Then, the feature extraction practice is done depending on power Local Directional Pattern (LDP) through statistical characteristics such as, mean, variance, and skewness [25].

Kaplan et al. (2020) proposed two dissimilar feature extraction approaches based on local binary pattern (LBP) for classifying the frequent brain tumor issues (nLBP and αLBP). The classification process was done by KNN, artificial neural network (ANN), random forest (RF), and linear discriminant analysis (LDA) with the feature matrices [26]. Raja and Rani (2020) developed a brain tumor classification method using a hybrid deep auto encoder with Bayesian fuzzy clustering (BFC) approach. In the pre-processing step, non-local mean filtering is performed for denoising purposes. Then, the BFC approach is used for the segmentation of brain tumors [27].

Ghahfarrokhi and Khodadadi (2020) proposed a computer-aided diagnosis (CAD) technique for the classification of tumors from MRI. A chaos theory is utilized for estimating the complexity measures such as Lyapunov exponent (LE), approximate entropy (ApEn), and fractal dimension (FD) while classifying the tumors. Here, Gray Level Co-occurrence Matrix (GLCM) and DWT-based features are extracted to identify the benign and malignant tumors. Subsequently, the extracted features are fed into three classifiers namely, SVM, KNN, and pattern net [28]. Deep learning (DL) using U-net is an automatic brain tumor segmentation method where the transfer learning through Vgg16 is employed for identifying the gliomas of brain tumor regions [29]. Shehab et al. (2020) proposed an automatic mechanism for tumor segmentation from brain images that depends on deep residual learning network (ResNet) to determine the gradient of deep neural network (DNN). Simulation is done with BRATS (2015) dataset to evaluate the superiority of the ResNet and it is observed that the computation time is lesser than the other DNN techniques [30].

Pareek et al. (2020) proposed a system to diagnose and classify the tumor. For experimentation, a set of 150 T1-weighted MRI brain images are utilized. Here, GLCM, DWT, and PCA are used for the feature extraction along with Kernel-SVM (KSVM) for classification. In this work, the area and volume of the tumor is also estimated to identify the different stages of the tumor [31]. Sajja and Kalluri (2020) proposed a technique where initially the MRI is enhanced by converting RGB into gray scale conversion. Further, FCM clustering is applied to segment the tumor regions in the brain MRI dataset. Later, LBP is used to extract the features and SVM is employed to classify the brain MRI as whether normal or abnormal [32].

Polepaka et al. (2019) proposed a method to label and trace the occurrence of tumor in brain MRI public dataset. The proposed method includes three phases namely, (i) preprocessing where the input image is filtered to diminish the noise, (ii) locating the tumor region from brain MRI, and (iii) classifying the tumor region by utilizing SVM [33]. Rad and Mosleh (2019) presented an automated brain tumor analysis system using a threshold-based segmentation method and tested with a Harvard medical school dataset containing 79 images. It combines the beta mixture model and learning automata (LA) for segmentation. Here, statistical features are considered and SVM, KNN, and decision tree (DT) strategies are applied as binary classifiers. The maximum accuracy about 98% is obtained using KSVM classifier with ten-fold cross-validation [34].

From the literature, it is inferred that the median filtering method is competent to denoise the MRI dataset. In addition to this, several potential edge detectors have been developed over the past decade and their efficiency should be investigated through common datasets. Previously, DWT was employed for feature extraction along with several state-of-the-art methods. K-means and Otsu's thresholding are the widely used techniques while considering the segmentation process. Altogether, ML methods are efficient in segmenting and classifying the tumor regions from brain MRI datasets. Hence, the present chapter is focused to validate the various state-of-the-art methods through a detailed investigation and comparative analysis.

7.3   PROPOSED METHODOLOGY

The proposed method of investigation is divided into five different phases as shown in Figure 7.2. They are: (i) image acquisition; (ii) preprocessing; (iii) segmentation; (iv) feature extraction; and (v) classification.

7.3.1   IMAGE ACQUISITION

The database used for the proposed work contains 240 brain MR images out of which 88 images are classified as normal and 152 images are known to be tumorous. Figure 7.3 presents the few sample images selected from the dataset. The entire MRI in the dataset is then resized and converted to gray scale images. Before segmentation, these images are subjected to various preprocessing steps for realizing maximum classification accuracy.

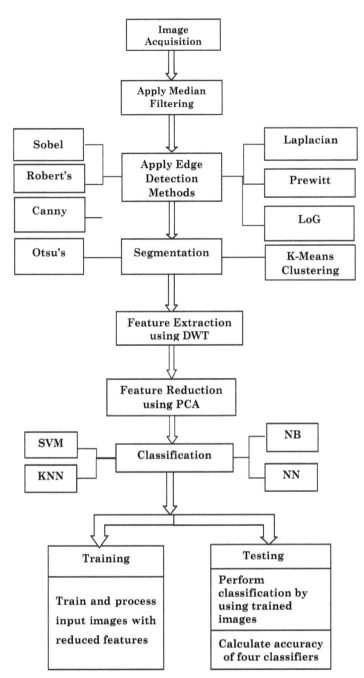

FIGURE 7.2   High level work flow of the proposed investigation.

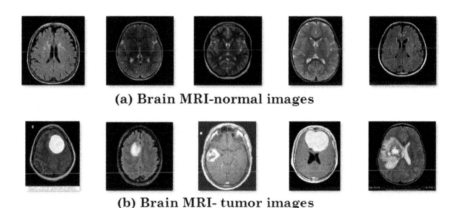

(a) Brain MRI-normal images

(b) Brain MRI- tumor images

FIGURE 7.3   Sample images from dataset. (a) without tumor (normal); (b) with tumor region.

7.3.2   PREPROCESSING

7.3.2.1   MEDIAN FILTERING

The collected MRI dataset contains noise because of multiple constraints and hence appropriate filtering is applied to remove the noise from source images. In this chapter, median filtering technique is employed that replaces the pixel intensity values with the median value of neighborhood pixels. Further, in this work 3×3 mask is employed and it provides effective results because of its excellent noise reduction capabilities.

7.3.2.2   EDGE DETECTION METHODS

The purpose of edge detection in image processing is to determine the areas of large intensities in an image. Here, in order to conclude an efficient edge detection method, 6 edge detection techniques are investigated in this chapter. The Sobel edge detector processes the image in X and Y directions and calculates the gradient of image intensity at every pixel. Sobel method detects the edges with maximum gradient value and the Sobel filter has two kernel (3×3) masks in horizontal (G_x) and vertical directions (G_y) [35]. The kernel mask used by Sobel operator is expressed as:

$$G_x = \begin{pmatrix} -1 & 0 & +1 \\ -2 & 0 & +2 \\ -1 & 0 & +1 \end{pmatrix}; \quad G_y = \begin{pmatrix} -1 & -2 & -1 \\ 0 & 0 & 0 \\ +1 & 2 & +1 \end{pmatrix}$$

Robert's edge detection operator uses the smallest kernel mask (2×2), and it detects the edges quickly with a significant amount of noise in the images. Kernel mask used by Robert's edge detection is expressed as:

$$G_x = \begin{pmatrix} -1 & 0 \\ 0 & -1 \end{pmatrix}; \quad G_y = \begin{pmatrix} 0 & -1 \\ 1 & 0 \end{pmatrix}$$

Prewitt edge detection also uses a 3×3 kernel mask with different coefficient values that are greater than Sobel operator. Hence, the Kernel mask used is given by:

$$G_x = \begin{pmatrix} -1 & -1 & -1 \\ 0 & 0 & 0 \\ +1 & +1 & +1 \end{pmatrix}; \quad G_y = \begin{pmatrix} -1 & 0 & +1 \\ -1 & 0 & +1 \\ -1 & 0 & +1 \end{pmatrix}$$

Canny edge detection uses a low pass filter before applying the Sobel operator and it detects the edges with minimum error [36]. The different steps employed in Canny edge detection can be summarized as follows: (i) Gaussian filter is used for filtering any noise that exists in the image, (ii). Determine the intensity gradient of image, (iii) Apply non-maximum suppression to remove the pixel values that are not to be a part of the image, and (iv) Detect the edges by hysteresis.

Laplacian edge detector is a second order derivative that detects the edges in a single pass. It highlights the region of MRI that has sharp intensity changes and Laplacian detector uses a single kernel as follows:

$$\begin{pmatrix} 0 & -1 & 0 \\ -1 & 4 & -1 \\ 0 & -1 & 0 \end{pmatrix}$$

LoG edge detector computes the edges by applying Gaussian filtering in combination with Laplacian detector. Gaussian filter is used for smoothening the image while the Laplacian is incorporated to compute the zero crossings. The input and output of this operator is a single gray level image.

7.3.3   SEGMENTATION

Segmentation is the process of dividing images into multiple regions to compute the region of interest (RoI) accurately. In this chapter, Otsu's Thresholding and K-means clustering techniques are used for the segmentation process to capture the tumor regions accurately.

7.3.3.1   OTSU'S THRESHOLDING

Otsu's thresholding is a clustering-based segmentation process. Here, in the first step probabilities and histogram of each pixel intensity level are calculated. Then, class means is calculated and iterated through all possible threshold values to determine the threshold that has maximum class variance value. By selecting the appropriate Graythresh levels, the tumor region from the image is efficiently extracted. After applying Otsu's Thresholding, images are converted to binary images containing '0's and '1's [37]. Figure 7.4 shows an example of the result of a noisy image after the implementation of Otsu's Thresholding. The steps involved for the segmentation process through Otsu's Thresholding are as follows: (i) Obtain the histogram of grayscale input image, (ii) Calculate the mean and variance from histogram, and then (iii) The pixel values are divided into two classes and calculate the within variance (v_w). The within variance function can be expressed as:

$$v_w = \sum w_i \times \sigma^2 \tag{1}$$

where; $w_i = \dfrac{No.\ of\ pixels\ in\ class\ i}{Total\ pixels}$; σ^2 denotes variance calculated from histogram. The between class variance (v_b) is computed as:

$$v_b = v_T - v_w \tag{2}$$

Then, the threshold value (T) is determined that has minimum v_b value and maximum v_w value.

7.3.3.2   K-MEANS CLUSTERING

K-means clustering is an unsupervised segmentation technique that extracts tumor regions from images. For this segmentation technique, in the first step the clusters (K) and their centers are initialized. Here, the K-value is defined as 4, since the images contain four colors. It will group the images into three regions and the region clustering is done by minimizing the Euclidean distance between the data and the corresponding cluster centroid. The tumor is extracted and displayed as one of the cluster results from most of the images.

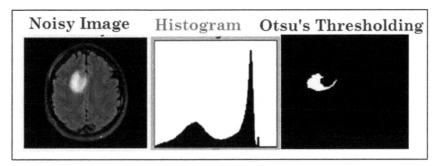

FIGURE 7.4   Example of Otsu's thresholding.

7.3.4   FEATURE EXTRACTION

The DWT along with GLCM is used for extracting the features from the segmented MRI. DWT is a powerful implementation of the WT that captures both frequency and location information. Suppose, if x (t) is a square-integrable function, then the continuous WT of x (t) relative to a given wavelet $\psi_{(t)}$ is defined as:

$$W_{\psi(a,b)} = \int_{-\infty}^{\infty} x(t)\psi_{a,b}(t)d(t) \tag{3}$$

where;

$$\psi_{a,b} = \frac{1}{\sqrt{a}}\psi(\frac{t-a}{b}) \tag{4}$$

The wavelet $\psi_{a,b}(t)$ is calculated from the mother wavelet $\psi(t)$ by the translation and dilation process. Where, 'a' is the dilation factor and 'b' is the translation parameter.

For feature extraction, DWT uses both low pass filters and high pass filters [38]. Then, it is down-sampled by a factor of 2 to obtain four sub-bands: LL, LH, HH, and HL. Each sub-band obtains wavelet coefficients in which LL obtains image approximation information, LH obtains horizontal features, HL obtains vertical features and HH obtains diagonal features. Further, GLCM is a second order statistical method that is used for finding a pair of pixels occurring in an image. In this chapter, 13 features are extracted by the contents of GLCM and the features are listed as follows:

1. **Mean:** It obtains mean pixel values of an input image.

$$Mean = \sum_i \sum_j i(P_{i,j})$$ (5)

where; $P_{i,j}$ is the $(i-j)^{th}$ entry of normalized GLCM.

2. **Variance:** It is the expectation of the squared deviation of a pixel from its mean value.

$$Variance = \sum_i \sum_j (1 - \mu)^2 P_{i,j}$$ (6)

where; $P_{i,j}$ is the $(i-j)^{th}$ entry of normalized GLCM; and μ is denoted as the mean of GLCM.

3. **Standard Deviation:** It is a measure of inhomogeneity and dispersion of pixel.

$$S.D = \sqrt{\sigma^2}$$ (7)

where; σ^2 denotes the variance of GLCM.

4. **Correlation:** It obtains mutual relationship or connections between two neighborhood pixels.

$$Correlation = \sum_i \sum_j P_{i,j} \frac{(i - \mu)(j - \mu)}{\sigma^2}$$ (8)

5. **Contrast:** It measures the amount of local variations present in an image.

$$Contrast = \sum_i \sum_j (i - j) P_{i,j}$$ (9)

6. **Entropy:** It measures the disorder or complexity of an image.

$$Entropy = \sum_i \sum_j P_{i,j} \log_2 P_{i,j}$$ (10)

7. **Angular Second Movement:** It measures the texture uniformity in pixel pair repetition.

$$ASM = \sum_i \sum_j P_{i,j}^2 \tag{11}$$

8. **Energy:** It gives the sum of square elements in GLCM.

$$Energy = \sqrt{ASM} \tag{12}$$

where; ASM denotes angular second movement.

9. **Homogeneity:** It measures image homogeneity and its value decreases when contrast value increases while the energy is kept constant.

$$Homogenity = \sum_i \sum_j \frac{1}{1+(i-j)^2} P_{i,j} \tag{13}$$

10. **Root Mean Square (RMS):** This value is obtained by taking the square of each pixel, and calculating the sum of those squares, and taking the square root.

$$RMS = \sqrt{\frac{\sum_{i=1}^{n}\left(X_i - \frac{\sum_{i=1}^{n} X_i}{n}\right)^2}{n}} \tag{14}$$

where; X_i denotes each pixel and 'n' is the total number of pixels.

11. **Inverse Difference Movement:** It is a measure of local homogeneity of an image.

$$IDM = \frac{\sum_i \sum_j P_{i,j}}{1+(i-j)^2} \tag{15}$$

12. **Kurtosis:** This describes the weight of random variables probability distribution.

$$Kurtosis = \frac{\sum_i \frac{X_i - \mu}{N}}{S^4} \tag{16}$$

where; X_i denotes each pixel, and 'S' represents the standard deviation.

13. Skewness: It is a measure of symmetry or lack of symmetry.

$$Skewness = \frac{\sum_i (x - \mu)^3}{\sigma^3} \tag{17}$$

where; 'x' is the pixel value and σ denotes the standard deviation.

7.3.5 FEATURE REDUCTION

The objective of feature reduction is to select meaningful features for the classification process from a set of features. PCA is one of the main feature reduction techniques that transform high dimensional features to low dimensional features. The reduced features obtained by PCA leads to better performance for classification of images into normal and abnormal classes.

7.3.6 CLASSIFICATION

The reduced features obtained from the feature reduction process are assigned as inputs for classification purposes. The proposed system uses four different classifiers namely, SVM, NB, KNN, and NN (FFNN and BPNN) for classifying the images into normal and abnormal classes. Firstly, the input dataset is divided into training and testing sets. In the training set, the input images are processed and trained with reduced feature vectors that are obtained from PCA. In the testing set, classification is performed based on training images and the classification accuracy of four classifiers is calculated.

7.3.6.1 SUPPORT VECTOR MACHINE (SVM)

SVM is a supervised classification technique for analyzing high dimensional data. Figure 7.5 presents the SVM classifier and it takes a set of images as input and classifies it into two classes (normal and abnormal) by cross validation. From training data, SVM obtains a model that classifies new images into appropriate classes. Subsequently, SVM will construct a set of hyperplanes that transforms low dimensional data to high dimensional data. Then, the SVM algorithm selects the hyperplane that has the largest margin to the nearest training data point of any class [39]. It is defined as:

$$f(y) = Z^T \phi(y) + b \tag{18}$$

where; Z and T are hyperplane parameters; b denotes bias; and (y) is a function used to map vector y into a higher-dimensional space.

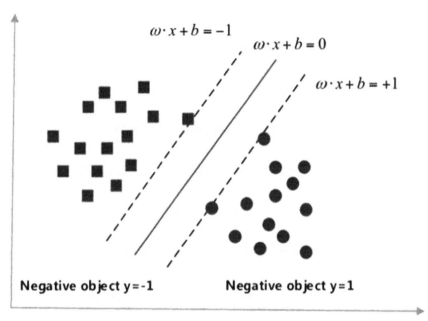

FIGURE 7.5 Support vector machine classifier. Maximum margin hyperplanes for a classifier that is trained with two sample classes.

7.3.6.2   K-NEAREST NEIGHBOR (KNN)

K-nearest neighbor (KNN) is a simplest supervised classification technique and it classifies the new images into the available category based on Euclidean distance. In the KNN algorithm, the value of 'K' (Nearest Neighbors) is a core deciding factor. It calculates the Euclidean distance to all data in the training set and sorts the distance in descending order. Then, the top 'K' values are selected from the sorted list. At last, the class of test points is defined based on the similarities for the selected value of 'K.' Figure 7.6 shows the essential steps involved in the working of KNN classifiers for classifying new data.

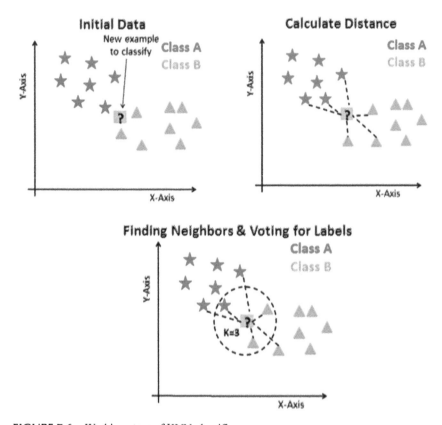

FIGURE 7.6   Working steps of KNN classifier.

7.3.6.3   NAÏVE BAYES

Naïve Bayes is a simple, powerful algorithm for classification. It is a supervised learning algorithm based on Bayes Theorem. This algorithm learns the probability of an object with certain features belonging to a particular class. It is easy to build and extremely useful in handling large datasets. Bayes theorem calculates posterior probability by:

$$P(c \mid x) = \frac{P(x \mid c)P(c)}{P(x)} \tag{19}$$

where; P(c|x) is the posterior probability of class; P(x|c) is the likelihood which is the probability of predictor given class; P(c) is the prior probability of class; P(x) is the prior probability of predictor.

7.3.6.4   NEURAL NETWORK CLASSIFIERS

7.3.6.4.1   Backpropagation Neural Network (BPNN) Classifier

Backpropagation is a supervised learning algorithm that is used for training neural networks (NN). A simple NN consists of three layers: Input Layer, Hidden Layer and Output Layer. The 13 features that are extracted are given as input to the NN in the input layer and corresponding classification results are obtained through the output layer. The output obtained is then compared with the original class and the mean square error is calculated. Then, these errors are backpropagated through the network and the weight is updated. This process is continued until minimum error is reached. Figure 7.7 shows the general BPNN architecture.

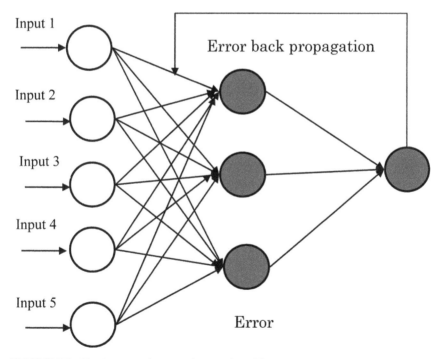

FIGURE 7.7   Backpropagation neural network architecture.

The output function (Sigmoid function) [40] used in ANN is given by:

$$O_j(\overline{x,w}) = \frac{1}{1+e^{A_i(\overline{x,w})}} \qquad (20)$$

where; $A_i^{(\overline{x,w})}$ is the activation function (weighted sum of input neuron (x_i) and corresponding weight, W_{ji}.

The error function is the sum of squared difference between target values t_k and the network output a_k and is given by:

$$E = \frac{1}{2}\sum_{k \in K}(a_k - t_k)^2 \qquad (21)$$

7.3.6.4.2   Feed Forward Neural Network (FFNN) Classifier

Since the information is moved in the forward direction, this network is called a feed forward network. The neurons in feed forward networks are organized as layers and these neurons are moving in only one direction (unidirectional). There are mainly three layers namely, Input Layer, Hidden Layer and Output Layers. The output of each layer is obtained from inputs that are multiplied with random weights. Since, the output value of this network is based on current input this network is called a static network. The output function (y) is given by:

$$y = f(w^T x + b) \qquad (22)$$

where; f denotes the activation function; w^T is the weight of each neuron; 'x' is the input neuron; and 'b' represents the bias.

FFNN is also called a memory less network because the output values are not related to the previous inputs [41]. Figure 7.8 shows the general architecture of FFNN with two hidden layers.

7.4   RESULTS AND DISCUSSION

7.4.1   RESULTS FOR PREPROCESSING PHASE

To evaluate the performance of four classifiers, a dataset which consists of 240 images is used. The dataset is included in two different folders as follows: A folder named as 'abnormal' which consists of images that have tumor and 'normal' which consists of non-tumor images. The sample images from the dataset with and without tumor are depicted in Figure

7.3(a) and (b), respectively. Filtering and edge detection methods are used as preprocessing steps. A 3×3 Median filtering is used for the removal of noise from these images. Sobel, Canny, Prewitt, Robert's, Laplacian, and LoG with sigma 3 are applied as edge detection methods. Results show that Canny edge detection performs better than other methods. The results obtained at the end of preprocessing steps are given in Figure 7.9. Figure 7.9(a) shows an input image, Figure 7.9(b) shows the image after applying median filter and the results corresponding to the application of various edge detection methods are shown in Figure 7.9(c).

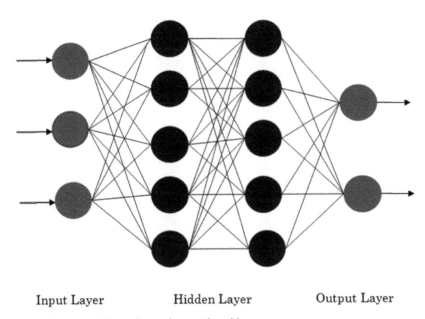

Input Layer   Hidden Layer   Output Layer

FIGURE 7.8   Feed forward neural network architecture.

7.4.2   RESULTS FOR SEGMENTATION TECHNIQUES

The proposed work uses two segmentation techniques Otsu's Thresholding and K means clustering. In K-means clustering, the value of K is considered as 4. That means, the image is segmented into 4 different regions and each region is represented by a different color. After the morphological operations are applied, dilation is used to avoid the unnecessary regions and Erosion is used to enhance the region. The results of two segmentation techniques are shown in Figure 7.10(a) and (b).

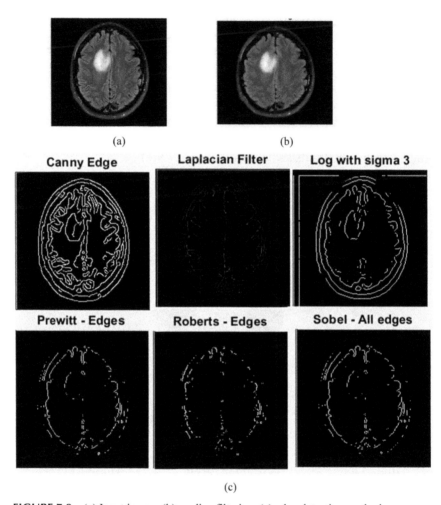

(a)   (b)

(c)

FIGURE 7.9   (a) Input image; (b) median filtering; (c) edge detection methods.

1. **K-Means Clustering:** After detecting the tumor regions from the
   MRI image by applying the appropriate segmentation techniques
   and the morphological operations, correlation is applied for deciding
   the tumor that is extracted can be correctly determined as a tumor.
   For this evaluation, normalized cross correlation has been performed
   using the function normxcor2(). In this process, the tumor extracted
   is compared with the original MRI image. If the result has high
   positive value (that is value close to 1), it is confirmed that the given

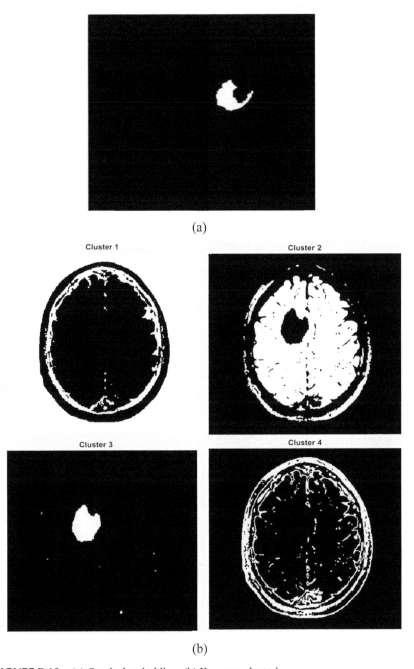

FIGURE 7.10   (a) Otsu's thresholding; (b) K-means clustering.

tumor is highly correlated with the original image. It means that the tumor is extracted efficiently with less noise. If the result has low positive value (that is value close to 1), then it is evident that the given tumor is less correlated with the original image. It means that the tumor is extracted with high noise (Table 7.1).

TABLE 7.1   Segmentation Techniques and Normalized Cross Correlation Value

Segmentation Techniques	Normalized Cross-Correlation Value
Otsu's	0.613
Local	0.616
Watershed	0.414
K means clustering	Cluster 1: 0.483
	Cluster 2: 0.155
	Cluster 3: 0.718
	Cluster 4: 0.348

From Figure 7.11, it is observed that K-Means Clustering in cluster 3 has a higher correlation value with 0.7184 than other segmentation techniques. The Otsu's, Local, and Watershed segmentation have the normalized cross correlation about 0.6131, 0.6162, and 0.4140 respectively.

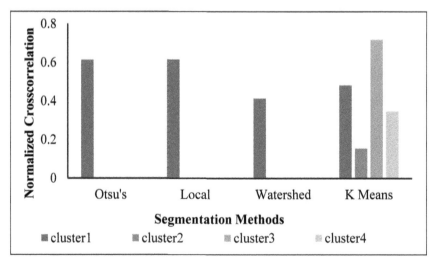

FIGURE 7.11   Comparison of segmentation methods using normalized cross correlation value.

7.4.3 RESULTS FOR FEATURE EXTRACTION PHASE

Feature Extraction is an important step in image processing. In this chapter, 13 features are extracted by using DWT along with GLCM. Totally 13 features are extracted and their corresponding values for the input image are shown in Table 7.2.

TABLE 7.2 Features Extracted Using DWT

SL. No.	Feature	Value
1.	Mean	0.0043
2.	S.D	0.0897
3.	Entropy	2.9207
4.	RMS	0.0898
5.	Variance	0.0079
6.	Smoothness	0.9412
7.	Kurtosis	16.2817
8.	Skewness	1.2917
9.	IDM	−0.3541
10.	Contrast	0.31868
11.	Correlation	0.0811
12.	Energy	0.7777
13.	Homogeneity	0.9354

Figure 7.12 shows the values of contrast, correlation, energy, entropy, and smoothness feature values of 10 images which are randomly selected.

7.4.4 RESULTS FOR CLASSIFICATION PHASE

In the classification phase, four different classifiers are used to classify images into normal and abnormal classes. For performance analysis, the confusion matrix is plotted and accuracy of classifiers are calculated. Accuracy is given by:

$$Accuracy = \frac{True\ positives + True\ negatives}{Total\ samples} \tag{23}$$

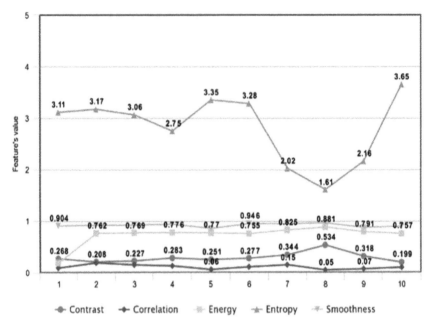

FIGURE 7.12   Five features of randomly selected images.

7.4.4.1   RESULTS FOR SVM CLASSIFIER

To classify, the images are first splitted into two: 70% for training 30% for testing. The SVM classifier correctly classifies 62 images as abnormal but misclassified 10 images as abnormal and obtains accuracy of 86.11%. The NB classifier also used 70% for training and 30% for testing. The NB classifier correctly classifies 54 images but misclassifies 18 images and obtains accuracy of 75%. The working of a KNN classifier mainly depends on the value of K and distance metric. Here, the value of K is chosen as 13(sqrt[n], where n is number of training samples) and distance metric used is Euclidean Distance. The KNN classifier also used 70% for training and 30% for testing. The KNN classifier correctly classifies 53 images but misclassified 19 images and obtains accuracy of 73.611%. In the Backpropagation NN classifier, we split images into 70% for training, 15% for validation and 15% for testing. The results of confusion matrix, Error histogram and overall performance plot of BPNN are shown in Figure 7.13. From the confusion matrix of BPNN, we can see that 176 images are correctly classified but 64 images are incorrectly classified and obtain an overall accuracy of 73.3%. The classification error occurred at intervals from 0.04629 to 0.8974 with 20 bins shown in Figure 7.13(b). The overall performance is shown in Figure 7.14. From the figure, it is observed that the best validation performance is 0.25717 at epoch 15.

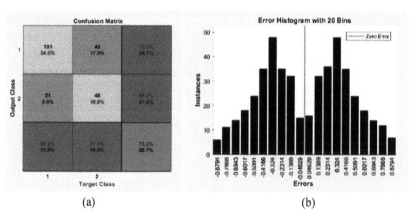

(a)   (b)

FIGURE 7.13   (a) Confusion matrix of BPNN; (b) error histogram.

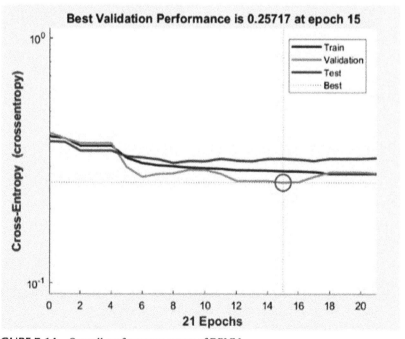

FIGURE 7.14   Overall performance curve of BPNN.

7.4.4.2 RESULTS FOR FEEDFORWARD NN

In FFNN classifier, the images are splitted into 70% for training, 15% for validation and 15% for testing. From confusion matrix of feedforward NN classifier, we can see that 179 images are correctly classified and 61 images are misclassified. Confusion matrices are shown in Figure 7.15(a). The error histogram occurred at intervals from 0.06809 to 0.9328 with 20 bins and overall performance of FFNN are shown in Figure 7.15(b) and Figure 7.16, respectively. From the performance curve, it is observed that best validation performance is 0.22728 at epoch 3.

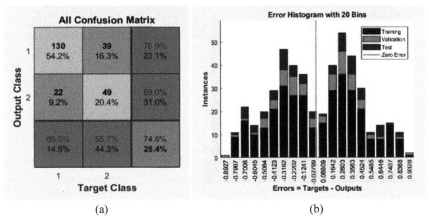

(a)   (b)

FIGURE 7.15   (a) Confusion matrix of FFNN; (b) error histogram.

7.4.4.3 COMPARISON OF CLASSIFIERS

The classifiers used for detection are compare based on Accuracy, Precision, Recall, and F1 Score.

1. **Accuracy:**

$$Accuracy = \frac{TP}{TP + TN + FP + FN} \qquad (24)$$

where; TP, TN, FP, FN are true positives, true negatives, false positives, and false negative values, accordingly.

The accuracy comparison of classifiers used by the proposed method with 70% training and 30% testing is shown in Figure 7.17.

The SVM classifier achieves the highest accuracy of about 86.11%. Other classifiers, like NB, KNN, BPNN, and FFNN obtain accuracies of 75%, 73.61%, 73.3% and 74.6%, accordingly.

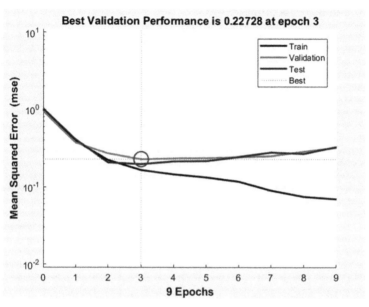

FIGURE 7.16   Overall performance curve of FFNN.

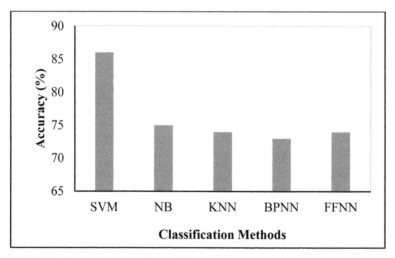

FIGURE 7.17   Accuracy comparison with 70% training and 30% testing.

The accuracy comparison of classifiers used with 60% training and 40% testing is shown in Figure 7.18. The SVM classifier has obtained the highest accuracy of 84%. Other classifiers, like NB, KNN, BPNN, and FFNN obtain accuracies of 73%, 62%, 70%, and 80% accordingly.

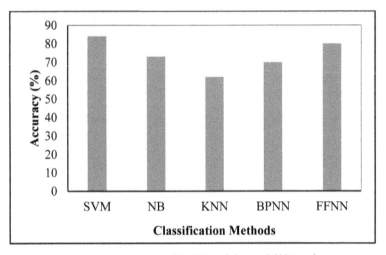

FIGURE 7.18　Accuracy comparison with 60% training and 40% testing.

The accuracy comparison of classifiers used with 65% training and 35% testing is shown in Figure 7.19. The SVM classifier obtained the highest accuracy 82%. Other classifiers, like NB, KNN, BPNN, and FFNN obtain accuracies of 67%, 69%, 71% and 76% accordingly.

The accuracy comparison of classifiers used with 75% training and 25% testing is shown in Figure 7.20. The SVM classifier obtained the highest accuracy 83%. Other classifiers, like NB, KNN, BPNN, and FFNN obtain accuracies of 70%, 71%, 72% and 75% accordingly.

2. **Precision:** It is defined as the rate of true positive value by total actual positive value. Where TP is the true positive; TN is the true negative; FP is the false positive; FN is the false negative:

$$Precision = \frac{TP}{TP + FP} \tag{25}$$

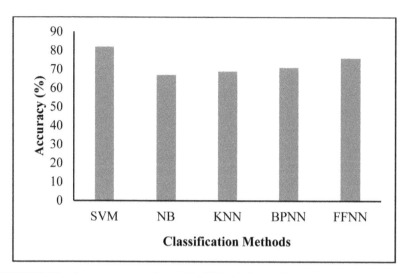

FIGURE 7.19   Accuracy comparison with 65% training and 35% testing.

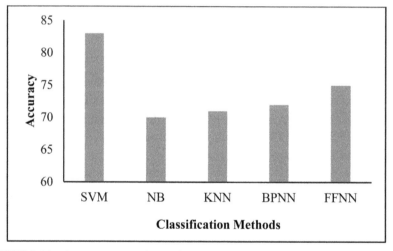

FIGURE 7.20   Accuracy comparison with 70% training and 30% testing.

The precision comparison of classifiers with 70% training and 30% testing are shown in Figure 7.21. The SVM classifier shows the highest precision of 100%. Other classifiers, like NB, KNN, BPNN, and FFNN obtain the accuracies of 87%, 82%, 75% and 76% accordingly.

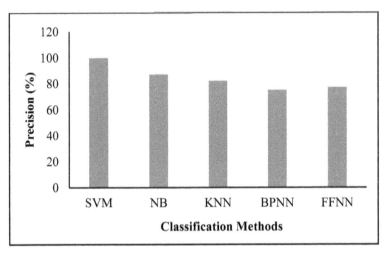

FIGURE 7.21   Precision comparison with 70% training and 30% testing.

3. **Recall:** It is the rate of true positive value by total predicted posi-
    tive value.

$$Recall = \frac{TP}{TP + FN} \qquad (26)$$

The recall comparison of classifiers with 70% training and 30%
testing are shown in Figure 7.22. The KNN classifier computes
the highest recall about 86.4%. Other classifiers, like NB, SVM,
BPNN, and FFNN obtain the accuracies of 84.3%, 86.1%, 86.2%
and 85.5% respectively.

4. **F1 Score:** It is defined as the weighted average of precision and
    recall.

$$F1score = \frac{2 \times precision \times recall}{precision + recall} \qquad (27)$$

The F1 score comparison of classifiers with 70% training and
30% testing sample images is shown in Figure 7.23. The SVM
classifier obtained the highest F1 score of about 92%. Other
classifiers, like NB, KNN, BPNN, and FFNN have reached the
accuracies of 85%, 84%, 81% and 80% respectively. From all
the comparative graphs of accuracy, precision, and F1 Score, it is
observed that the SVM has obtained the highest value while KNN
achieves the highest value of Recall only.

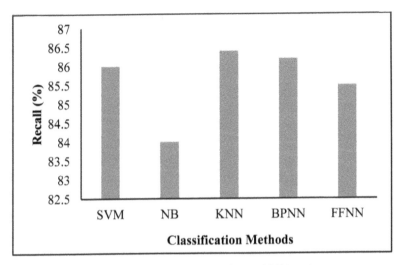

FIGURE 7.22   Recall comparison with 70% training and 30% testing.

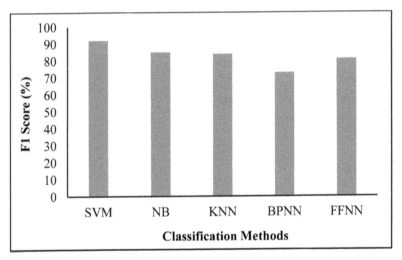

FIGURE 7.23   F1 score comparison with 70% training and 30% testing.

7.5 CONCLUSION

In this presented chapter, an investigation of various ML methods for detection and classification of tumor and non-tumor regions in MRI is done with state-of-the-art methods. After preprocessing the MRI dataset, edges are detected and tumor regions are segmented using Otsu's Thresholding and K-Means Clustering. Among these ED techniques, Canny edge detection detects the edges more accurately than other methods. DWT is used for feature extraction and 13 features are extracted for classification process. SVM, NB, KNN, and NN (BPNN and FFNN) classifiers are used for the classification process. Out of which, the SVM classifier with Otsu's thresholding has better performance with 86.11% accuracy. In the future, state-of-the-art feature extraction techniques can be investigated to identify their influence on the classifier performance. Alternatively, advanced DL techniques also can be modified and applied for determining the brain disorders.

KEYWORDS

- artificial neural network
- brain tumor
- classification techniques
- convolution neural network
- edge detection
- extreme learning machines
- feature extraction

REFERENCES

1. Ekaterina, F., Konstantina, K., Susanne, U., Nicolás, G. N., Sebastian, U., Elisabeth, J. R., Luca, R., et al., (2020). Single-cell mapping of human brain cancer reveals tumor-specific instruction of tissue-invading leukocytes. *Cell, 181*(7), 1626–1642.e20. doi: 10.1016/j.cell.2020.04.055.
2. Bibal, B. J. V., & Venifa, M. G., (2020). Segmentation and classification of tumor regions from brain magnetic resonance images by neural network-based technique. In: *Book: Handbook of Artificial Intelligence in Biomedical Engineering* (1st edn.). Chapter: 20. Apple Academic press, Taylor & Francis Group. ISBN: 9781771889209.
3. Anjali, W., Anuj, B., & Vivek, S. V., (2019). A review on brain tumor segmentation of MRI images. *Magnetic Resonance Imaging, 61*, 247–259. doi: 10.1016/j.mri.2019.05.043.

4. Hosseinzadeh, M., Salmani, S., & Majles, A. M. H., (2019). Interferometric optical testing to discriminate benign and malignant brain tumors. *Journal of Photochemistry and Photobiology B: Biology, 199,* 111590. doi: 10.1016/j.jphotobiol.2019.111590.

5. Jakub, N., Pablo, R. L., Michal, M., Bobek-Billewicz, B., Pawel, W., Maksym, W., Michal, K., et al., (2020). Fully-automated deep learning-powered system for DCE-MRI analysis of brain tumors. *Artificial in Intelligence Medicine, 102.* DOI: 10.1016/j.artmed.2019.101769.

6. Anwar, S., Javed, I. B., Abdul, W. K., Imran, A., Abdullah, K., Asfandyar, K., & Arshad, K., (2020). Comparative analysis of median filter and its variants for removal of impulse noise from gray scale images. *Journal of King Saud University - Computer and Information Sciences.* doi: 10.1016/j.jksuci.2020.03.007.

7. Bidyut, B. C., & Bhabatosh, C. (1984). The equivalence of best plane fit gradient with Robert's, Prewitt's and Sobel's gradient for edge detection and a 4-neighbour gradient with useful properties. *Signal Processing, 6*(2). doi: 10.1016/0165-1684(84)90015-X.

8. Yingchao, M., Zhongping, Z., Huaqiang, Y., & Tao, M., (2018). Automatic detection of particle size distribution by image analysis based on local adaptive canny edge detection and modified circular Hough transform. *Micron, 106,* 34–41. doi: 10.1016/j.micron.2017.12.002.

9. Phillip, A. M., & Jeffrey, J. R., (2009). Chapter 19 - gradient and Laplacian edge detection. In: Al-Bovik, (ed.), *The Essential Guide to Image Processing* (pp. 495–524). Academic Press. doi: 10.1016/B978-0-12-374457-9.00019-6.

10. Ta Yang, G., Shafriza, N. B., Haniza, Y., Muhammad, J. A. S., & Fathinul, S. A. S., (2018). Performance analysis of image thresholding: Otsu technique. *Measurement, 114,* 298–307, doi: 10.1016/j.measurement.2017.09.052.

11. Kai, T., Jiuhao, L., Jiefeng, Z., Asenso, E., & Lina, Z., (2019). Segmentation of tomato leaf images based on adaptive clustering number of k-means algorithm. *Computers and Electronics in Agriculture, 165.* doi: 10.1016/j.compag.2019.104962.

12. Carlos, A., Alfredo, V., & Enrique, R., (2012). Classification of human brain tumors from MRS data using discrete wavelet transform and Bayesian neural networks. *Expert Systems with Applications, 39*(5), 5223–5232. doi: 10.1016/j.eswa.2011.11.017.

13. Xianchao, X., Ying, Y., Lingchen, K., & Wanquan, L., (2020). Laplacian regularized robust principal component analysis for process monitoring. *Journal of Process Control, 92,* 212–219. doi: 10.1016/j.jprocont.2020.06.011.

14. Norbert, B., Thomas, B., Bernd, F. M. R., Rupert, R., Rolf, K., Christoph, K., & Jürgen, P., (2012). Identification of primary tumors of brain metastases by Raman imaging and support vector machines. *Chemometrics and Intelligent Laboratory Systems, 117,* 224–232. doi: 10.1016/j.chemolab.2012.02.008.

15. Caitlin, C., (2020). Parameter tuning naïve bayes for automatic patent classification. *World Patent Information, 61,* 101968. doi: 10.1016/j.wpi.2020.101968.

16. Albert, C., Alessandro, S., Giorgio, R., Samuel, B., Maria, G. S., Giovanni, P., Massimo, I., & Anthony, Y., (2019). K-nearest neighbor driving active contours to delineate biological tumor volumes. *Engineering Applications of Artificial Intelligence, 81,* 133–144. doi: 10.1016/j.engappai.2019.02.005.

17. Chunrui, Z., Xianhong, Z., & Yazhou, Z., (2018). Dynamic properties of feed-forward neural networks and application in contrast enhancement for image. *Chaos, Solitons & Fractals, 114,* 281–290. doi:10.1016/j.chaos.2018.07.016.

18. Telrandhe, S. R., Pimpalkar, A., & Kendhe, A., (2016). Detection of brain tumor from MRI images by using segmentation & SVM. *World Conference on Futuristic Trends in Research and Innovation for Social Welfare.*

19. Rashid, M. H. O., Mamun, M. A., et al., (2018). Brain tumor detection using anisotropic filtering, SVM classifier and morphological operation from MR images. *International Conference on Computer, Communication, Chemical, Materials and Electronic Engineering.*

20. Zuliani, Z., (2019). Detection of brain tumor and extraction of features in MRI images using K-means clustering and morphological operations. In: *2019 IEEE International Conference on Automatic Control and Intelligent Systems (I2CACIS).*

21. Zaw, H. T., Maneerat, N., & Win, K. Y., (2019). Brain tumor detection based on naïve bayes classification. In: *5th International Conference on Engineering, Applied Sciences and Technology (ICEAST).*

22. Ahmed, K., Med, B. M., et al., (2019). Detection of brain tumor in medical images. In: *3rd International Conference on Signals, Circuits & Systems (IEEE SCS'09).*

23. Fatih, Ö., Eser, S., & Derya, A., (2020). An expert system for brain tumor detection: Fuzzy C-means with super resolution and convolutional neural network with extreme learning machine. *Medical Hypotheses, 134,* 109433. doi: 10.1016/j.mehy.2019.109433.

24. Sharan, K., (2020). Optimization driven deep convolution neural network for brain tumor classification. *Biocybernetics and Biomedical Engineering.* doi: 10.1016/j.bbe.2020.05.009.

25. Shubhangi, N., Akshay, D., Subrahmanyam, M., & Srivatsava, N., (2020). RescueNet: An unpaired GAN for brain tumor segmentation. *Biomedical Signal Processing and Control, 55.* doi: 10.1016/j.bspc.2019.101641.

26. Kaplan, K., Yılmaz, K., Melih, K., & Metin, E. H., (2020). Brain tumor classification using modified local binary patterns (LBP) feature extraction methods. *Medical Hypotheses, 139,* 109696. doi: 10.1016/j.mehy.2020.109696.

27. Siva, R. P. M., & Antony, V. R., (2020). Brain tumor classification using a hybrid deep autoencoder with Bayesian fuzzy clustering-based segmentation approach. *Biocybernetics and Biomedical Engineering, 40*(1). doi: 10.1016/j.bbe.2020.01.006.

28. Sepehr, S. G., & Hamed, K., (2020). Human brain tumor diagnosis using the combination of the complexity measures and texture features through magnetic resonance image. *Biomedical Signal Processing and Control, 61.* doi: 10.1016/j.bspc.2020.102025.

29. Mohamed, A. N., & Jamal, D. M., (2020). Brain tumor segmentation and grading of lower-grade glioma using deep learning in MRI images. *Computers in Biology and Medicine, 121.* doi: 10.1016/j.compbiomed.2020.103758.

30. Lamia, H. S., Omar, M. F., Safa, M. G., Mohamed, S., & El-Mahallawy, (2020). An efficient brain tumor image segmentation based on deep residual networks (ResNets). *Journal of King Saud University - Engineering Sciences.* doi: 10.1016/j.jksues.2020.06.001.

31. Pareek, M., Jha, C. K., & Mukherjee, S., (2020). Brain tumor classification from MRI images and calculation of tumor area. In: Pant, M., Sharma, T., Verma, O., Singla, R., & Sikander, A., (eds.), *Soft Computing: Theories and Applications. Advances in Intelligent Systems and Computing* (Vol. 1053). Springer, Singapore.

32. Sajja, V., & Kalluri, H. K., (2020). Brain tumor segmentation using fuzzy C-means and tumor grade classification using SVM. In: Fiaidhi, J., Bhattacharyya, D., & Rao, N., (eds.), *Smart Technologies in Data Science and Communication. Lecture Notes in Networks and Systems* (Vol. 105). Springer, Singapore.

33. Polepaka, S., Srinivasa, R. C., & Chandra, M. M., (2019). A brain tumor: Localization using bounding box and classification Using SVM. In: Saini, H., Singh, R., Patel, V., Santhi, K., & Ranganayakulu, S., (eds.), *Innovations in Electronics and Communication Engineering. Lecture Notes in Networks and Systems* (Vol. 33). Springer, Singapore.

34. Edalati-Rad, A., & Mosleh, M., (2019). Improving brain tumor diagnosis using MRI segmentation based on collaboration of beta mixture model and learning automata. *Arab J. Sci. Eng., 44,* 2945–2957. doi: 10.1007/s13369-018-3320-1.

35. Hafiza, H. T., Syed, S. A., & Haroon, R., (2015). Tumor detection through image processing using MRI. *International Journal of Scientific & Engineering Research, 6*(2).

36. Chithambaram, T., & Perumal, K., (2017). Brain tumor detection and segmentation in MRI images using neural network. *International Journal of Advanced Research in Computer Science and Software Engineering, 7*(3).

37. Shahriar, S. T. M., Misbah, U. H., et al., (2019). Development of automated brain tumor identification using MRI images. *International Conference on Electrical, Computer and Communication Engineering (ECCE).*

38. Shrutika, S., (2017). Implementation of image processing for detection of brain tumors. *International Conference on Intelligent Computing and Control Systems.*

39. Nilesh, B. B., Arun, K. R., et al., (2017). Image analysis for MRI based brain tumor detection and feature extraction using biologically inspired BWT and SVM. *International Journal of Biomedical Imaging, 2017.*

40. Heung, S., (2017). Chapter 1 - an introduction to neural networks and deep learning. In: Kevin, Z. S., Hayit, G., & Dinggang, S., (eds.), *Deep Learning for Medical Image Analysis* (pp. 3–24). Academic Press. doi: 10.1016/B978-0-12-810408-8.00002-X.

41. Senthilkumar, M., (2010). In: Gulrajani, M. L., (ed.), *5-Use of Artificial Neural Networks (ANNs) in Color Measurement* (pp. 125–146). In Woodhead Publishing Series in Textiles, Color Measurement, Woodhead Publishing. doi: 10.1533/97808 57090195.1.125.

CHAPTER 8

TUMOR DETECTION BASED ON 3D SEGMENTATION USING REGION OF INTEREST

T. M. RAJESH, S. G. SHAILA, and LAVANYA B. KOPPAL

Department of Computer Science, Dayananda Sagar University, Bangalore, Karnataka, India, E-mail: shaila-cse@dsu.edu.in (S. G. Shaila)

ABSTRACT

The chapter addressed two complex issues segmenting 2D slices of CT or MRI and view segmented 2D slices in 3D viewer. For segmenting 2D slices, we used the watershed algorithm. The 3D viewer is used for better visualization, which gives more information about segmented tumors. For better segmentation, included Gradient technique with the watershed algorithm. Segmented series of 2D images are viewed in 3D viewer. Experimental result shows viewing segmented 2D slices in 3D viewer. In our proposed methodology, we have shown the essence and feasibility of an automated tumor segmentation method for both CT and MRI images, and a simple model consists of the watershed algorithm is used to segment the tumor. The method segments a series of slices that consists of tumors and can be viewed using the 3D viewer technique and has been validated on five clinical MRI datasets, which consists of a minimum of 20–25 slices. The end results show a promising result.

8.1 INTRODUCTION

The field of medical image processing refers to processing MRI or CT images. Medical image processing plays vital role in the current era, where in automation is at most demand in the society. Detection of fractures, cancer, and tumor belongs to the club of major domains in medical image processing. In the good old era, it was very difficult to trace and detect damaged parts of the body manually. As technology evolved MRI, CT scans emerged and helped the doctors, radiologists, analyzers, and researcher's groups a lot in diagnosing process. For automatic diagnosis well supporting application and software plays a vital role in the background. Complicated algorithms involve too many methodologies and iterations in the process, which eventually results the radiologists or researchers to end up in dilemma. So basically, an algorithm or the process of the application should be reliable, accurate, robust, and less time consuming in terms of generating the end results. The smooth and simple process is the need of an hour. In this work we have tried to give a solution for simple process to segment the tumor in the brain and giving a 3D look to the segmented tumor. A segmented tumor 3D model is better in all terms compared to MRI dataset for analysis and diagnosis purpose.

After lots of rigorous survey, it is found the watershed algorithm is best suitable for segmentation of the tumor in the MRI data. Segmentation is the process of extracting the wanted data based on region of interest (RoI). Also, in other terms extracting the meaningful information from the unwanted region in the image is nothing but segmentation. This process makes the researcher job easier in analysis phase or feature extraction phase. The process of segmentation is mainly used to locate the tumor and its features like size, boundary, and curve in the image. A group of segmented images makes a counter or a voxel. A voxel is representation of a volume in the three-dimensional space. These voxels will be commonly used to visualize and analysis purpose. So, creating the visualization of segmented tumor is challenging task in this process. Example of voxel is as shown in Figure 8.1.

In this work, the main reason of using 3D Image Segmentation is to slice a 3D image into regions that can be considered as same as the real time visualization with respect to a given insight such as region based and to detect the tumor portion. The chapter is followed by Literature survey in Section 8.2 and followed by proposed methodology in Section 8.3 and results and discussion in Section 8.4 followed by conclusion in the end Section 8.5.

8.2   LITERATURE SURVEY

Julien Mille et al. [1] proposed a model which concentrates on segmentation and reconstruction of 3D vascular trees. The topological relationships

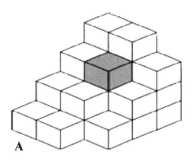

A series of Voxels in a stack. A   A dataset of Voxels for a macromolecule
single Voxel is highlighted.

FIGURE 8.1   Voxel representation and data set of voxels for macromolecule.

Source: A: Image by Vossman; M. W. Toews. https://creativecommons.org/licenses/by-
sa/2.5/

B: Image by Vossman. https://creativecommons.org/licenses/by-sa/2.5/deed.it

between the segments of the deformable tree are modeled. The results got on CTA images of the carotid bifurcation are presented in their work. This will help in posterior interactions and analysis such as measuring lengths or diameters of the vessels. Sheng You et al. [2] have provided a novel approach of applying of principal surfaces for the propagation of contours in 4D-CT study. In reference to 3D-CT scans Regions of Interest are manually demarcated. In the target 3D-CT phase the edges are detected on all the slices. Preliminary results are presented on the approach. There is a necessity to incorporate prior shape information into the proposed procedure to make principle surfaces smoother and steadier with the prior shapes. Tao Wang et al. [3] proposed a novel approach called active contour model to deal with limitations of insufficient capture range and poor convergence of concavities. It provides an improvement over methods such as gradient vector flow, boundary vector flow and magneto static active contour of three different sets of experiments. Rajeev et al. [4] proposed a brain tumor segmentation method which was developed and validated based on segmentation of 2D and 3D MRI Data. This method can segment a tumor only when the desired parameters are set properly. Unlike other methods it does not require any initialization inside the tumor. Wang et al. [5] reconstructs three dimensional (3D) models of human body by using CT slices, digital images which helps to accurately find the locations of pathological formations such as tumors. It depicts the implementing fundamental limitations in 3D medical image reconstruction of medical imaging such as marching cubes (MC) algorithms, usual rendering technique, etc., and designing software for reconstructing 3D image from a set of CT images. The results show that the reconstruction could help to reconstruct series of 2D segmented binary images and display the 3D image of the target object. Przemyslaw et al. [6] provides a model for 3D image segmentation and reconstruction. It has been considered with the aim to be implemented on a computer cluster or a multi-core platform. The features include almost absolute independence between the processes participating in the segmentation task and providing work as equal as possible for all the participants. Vasupradha et al. [7] proposed an enhanced Darwinian particle swarm optimization (EDPSO) for automated tumor segmentation which surmounts the drawback of existing particle swarm optimization (PSO). Tracking algorithm is used for pre-processing the MRI images and Gaussian filter. The segmentation is used using the PSO. Javeria et al. [8] proposed an automated method for detection and classification of tumors by using at the image and lesions level. The candidate lesions are segmented incorporating different techniques. Grade identification is done by selecting a hybrid feature set and three variants of SVM are tested. Shrutika et al. [9] describes the detection of the brain tumor by the method of thresholding. The proposed method can be competently applied to detect and extract the brain tumor from MRI images obtained from patient's database. Manisha et al. [10] proposed a method for detection of brain tumor using Sobel edge detection method. The MRI images have been undergone pre-processing like smoothing using median filter. An appropriate method to find the threshold value using standard deviation is carried out. At last Sobel edge detector was used to find the border of the tumor region. Annisa et al. [11] proposed a method where the segmentation of brain is done by providing a mark using watershed method to the area of the brain and areas outside the brain by clearing skull with cropping method. Here, 14 brain tumor MRI images are used. The segmentation results are compared brain tumors area and brain tissues area. This system obtained the calculation of tumor area has an average error of 10%. Minal et al. [12] proposed a method for detection and segmentation of brain tumor from T1-weighted and fluid-attenuated inversion recovery (IR) brain images. Fractional Sobel filter is used for pre-processing the images which improves the segmentation results. Classification is done using SVM. Mircea et al. [13] proposed a methodology for detection and classification of brain tumor images from MRI using numerous wavelets transform and SVM (support vector machine). Shahriar et al. [14] proposed an automated method where MRI gray scale images were used for detection of tumor. After applying many pre-processing techniques, OTSU segmentation method is used instead of color segmentation.

8.3   PROPOSED APPROACH

The proposed methodology segments 2D slices of CT or MRI using watershed algorithm. The segmented 2D slices of CT or MRI are stored in a folder, then for the 3D visualization purpose picture viewer is used. The work is carried out in two steps:

- Segmenting 2D slices of CT or MRI using watershed segmentation algorithm; and
- 3D visualization is done using picture viewer.

Figure 8.2 shows the overall block diagram of 3D segmentation based on RoI.

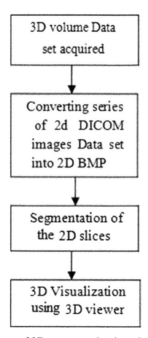

FIGURE 8.2   Proposed system of 3D segmentation based on region of interest (RoI).

8.3.1   3D VOLUME DATA SET ACQUIRED

Radiologists, Practitioners, and Researchers are found of using DCM images which are generated by 3-D software application called magnetic resonance imaging (MRI) scans from various angles to see the damaged portion of the body from normal view. The grayscale image is subjected as the input to a segmentation procedure, for example the output of MRI scan. The probable output, or "segmented images," contains the labels that classify the input grayscale volume. Figure 8.3 represents 3D data sets of the brain, along with the original 2D slices used to apply the segmentation.

8.3.2   CONVERTING SERIES OF 2D DICOM IMAGES DATA SET INTO 2D BMP SLICES

After collecting the dataset, the volume which consists of series of slices are converted into .jpeg or .bmp format which are in DICOM images (.dcm) for easier process. After converting the formats those slices are considered as 2D images, further segmentation technique using watershed algorithm will have implemented to separate the tumor portion from all the tumor contained slices.

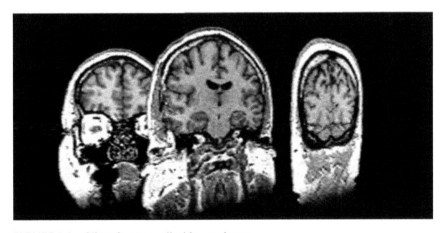

FIGURE 8.3   Slices from a medical image dataset.

8.3.3   SEGMENTATION OF ALL 2D SLICES

One of the basic steps in scientific and medical image processing is automatic segmentation. The process of segmentation involves cropping of the unwanted portion of the image. In cropping, unwanted region will be skimmed off with the help of pixels. Neighboring pixels have the concepts of 4-neighboring pixels, 8-neighboring pixels, and M-neighboring pixels. These concepts will be used to pass on or extract the information from the neighboring pixels. First the object or the region of the interest should be identified. In our work we have used watershed algorithm to identify the boundary region of the tumor. The identified boundary region pixels are considered as the foreground and remaining pixels are considered as the foreground. In automated segmentation process foreground region pixels will be subtracted from background region pixels. Also, not every algorithm is capable of detecting the tumors like brain tumor or kidney tumor. Here, in this work we have been used watershed techniques for detecting the tumors. Automated cropping for segmentation and then, for 3D visualization we used the concept picture viewer, which gives 3D visualization and accurate segmented output.

8.3.3.1   SEGMENTATION USING WATERSHED ALGORITHM

Types of connecting points in a topographic interpretation:

- Points belonging to a regional minimum;
- Points at which a drop of water would fall to a single minimum;
- Points at which a drop of water would be equally likely to fall to more than one minimum;
- The main purpose is to find watershed boundary lines as shown in Figure 8.4.

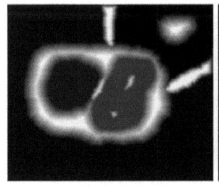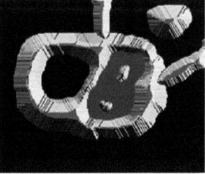

FIGURE 8.4   Example of watershed lines.

Source: Image from https://www.pianshen.com/article/86001663083/

The basic idea is as shown in Figure 8.5:

- Suppose when it is raining and how the hilly regions are covered with water leaving out the hilly region portion. The algorithms used in this works covers up the tumor portion in the MRI.
- Suppose the water flow is too high and how the dam wall is used to stop the outflow. The proposed algorithm dam wall can be considered as the boundary line of the tumor (the divide lines or watershed lines).

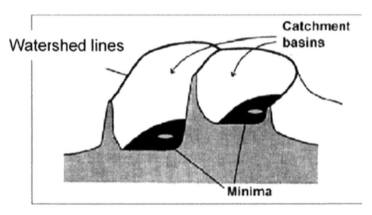

FIGURE 8.5   Example for how watershed works.

8.3.3.2   WORKING PROCEDURE OF WATERSHED SEGMENTATION ALGORITHM

1. Create the Kernel which moves all over the image to capture the required features of the tumor. The outcome should be something like, lowest possible value is zero or black pixels and highest possible value is 1 or white pixel.

     These identified pixels help us in forming the basis for initial watersheds.
2. For each pixel value 'a':
     For each group of pixels of volume 'A'
     If its adjacent to RoI, add 'A' to that region;
     Else if,
          Adjacent to more than one RoI, mark as boundary line
     Else,
          Start a new RoI

8.3.3.3 ALGORITHM

1. P1, P2, P3,...Pn: Min(RoI)of an image g(x,y).
2. C(Pi): RoI + Min Pi.
3. S[n]: S€(x,y) ∀ T[n] < g(x,y) < n.
4. T[n]=(x,y) * Log(x,y) < n
5. T[n]: S € g(f(x, y)) < g(f(x, y)) = n.
6. n = min + 1 to n = max + 1.
7. If the coordinates T[n] < g(f(x,y)) = n = 0 else 1. (0 = Black & 1 = white)

8.3.3.4 D VISUALIZATION USING 3D VIEWER

For 3D visualization of all the 2D slices a procedure called 3D viewer can be used. All 2D slices are need to be arranged one before the other according to their positions and then giving the time interval between each of the slices for the moment which appears as the 3D view. In segmented slices the original information will be lost, and cannot be able to reconstruct the same segmented images, so for this purpose used this 3D viewer technique.

8.4 RESULTS AND DISCUSSIONS

The method segments a series of slices which consists of tumor and can be viewed using 3D viewer technique and has been validated on 5 clinical MRI datasets, each dataset consists of minimum 20–25 slices. The results are presented using visual C++ and MATLAB and same has been represented in Sections 8.4.1 and 8.4.2.

8.4.1 RESULTS USING VISUAL C++

We have used visual C++ framework for the analysis. We followed the process has represented below:

- Segmented single BMP and JPEG slice;
- Used Watershed algorithm for segmentation;
- Manually segmented each slice and stored those segmented slices in one folder;

- Displayed those segmented slices using the concept picture viewer, which can be viewed like 3D image only.

In segmented slices the original information will be lost, so for this purpose 3D reconstruction method cannot be done.

The 2D slices which are acquired from the CT or MRI are stored in the folder and displayed the same image for the analysis of the original image as shown in Figure 8.6.

FIGURE 8.6   Original image.

After reading and displaying the original image, segmentation is applied using the watershed segmentation algorithm as shown in Figure 8.7.

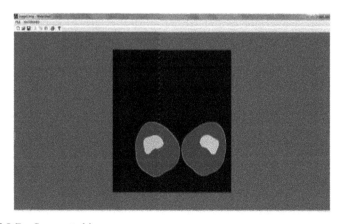

FIGURE 8.7   Segmented image.

All the segmented slices are stored in one folder as shown in Figure 8.8.

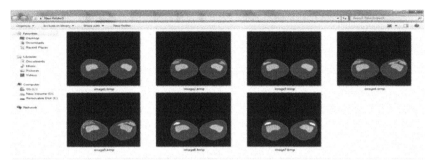

FIGURE 8.8   Segmented images stored in one folder.

All the segmented slices are viewed in the 3D viewer for 3D visualization. This gives specific information about the tumor which is there in tumor effected slices and is represented in Figure 8.9.

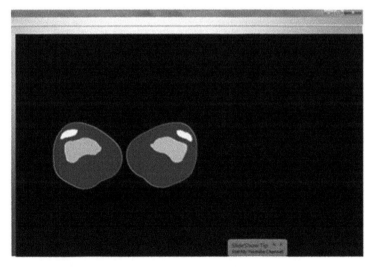

FIGURE 8.9   Picture viewer.

8.4.2   RESULTS USING MATLAB

1. Segmented series of slices (DICOM, BMP, and JPEG) are taken automatically and stored in one folder:

   i. Used gradient technique for segmentation;

   ii. Worked even for both smaller and larger tumors.

 2. For displaying segmented slices, techniques are used similarly as in VC++, but here difference is, slices can be viewed in 3D format in all X, Y, and Z axis.

The 2D slices which are acquired from the CT or MRI are stored in the folder as shown in Figure 8.10. The stored slices are segmented using watershed segmentation algorithm and stored in another folder as shown in Figure 8.11. The stored segmented slices are viewed in the 3D viewer for better 3D visualization as shown in Figure 8.12. The performance of the approach demonstrated by extensive experimental results confirms Segmented 2D slices are viewed in 3D viewer will give clear and detail information about the segmented tumor portion. Also Segmenting 2D slices and then viewing in 3D viewer achieves the low computation cost and the minimum user interaction. This is represented in Figure 8.13.

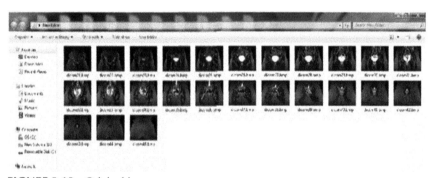

FIGURE 8.10 Original images.

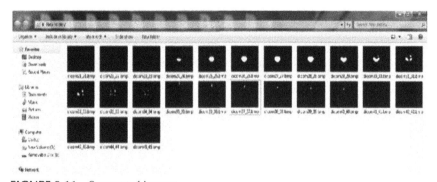

FIGURE 8.11 Segmented images.

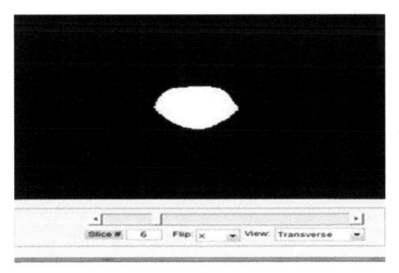

FIGURE 8.12   Picture viewer of segmented images.

TABLE 8.1   Accuracy Analysis based on Number of Slices/Volume

SL. No.	Number of Slices	Segmentation Accuracy (%)
1.	500	93
2.	400	93.5
3.	300	95
4.	200	98
5.	100	100
6.	50	100

8.5   CONCLUSION AND FUTURE WORK

The proposed work addressed two complex issues segmenting CT or MRI dataset volume into 2D slices and view segmented 2D slices in 3D viewer. 3D viewer gives comfortable zone for the analyzer to have a 360' view over the RoI. For segmenting 2D slices we have used watershed algorithm. 3D viewer used for better visualization which gives more information about segmented tumors. For better segmentation we have also included Gradient technique with watershed algorithm. Experimental result shows viewing segmented 2D slices in 3D viewer and is giving promising results. In this work we have tried to give an automation touch for the manual procedure which has been carried out since ages. Also, the automated work consists of minimum algorithms and procedure which eventually helps the analyzer or the researcher to extract features in a minimum time. Basically, the proposed work is built on simple knowledge model. In our future work we will be applying the segmentation process directly to 3D volumes using rendering technique. In the current procedure 3D volume has been converted into slices and then segmentation algorithm is applied on the individual slices. Later these slices are given the 3D touch but keeping min time interval among the slices. This process can be improved and can be made much easier, if we apply the segmentation process directly to 3D volumes or voxels.

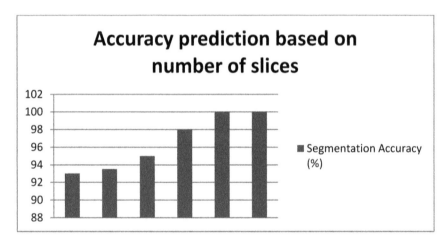

FIGURE 8.13   Accuracy prediction based on the number of slices/volumes.

KEYWORDS

- **3D view**
- **magnetic resonance imaging**
- **marching cubes**
- **region of interest**
- **segmentation**
- **tumor**
- **watershed algorithm**

REFERENCES

1. Julien, M., & Laurent, D. C., (2010). 3D CTA image segmentation with a generalized cylinder-based tree model. *IEEE International Symposium on Biomedical Imaging: From Nano to Macro.*

2. Sheng, Y., Ataer-Cansizoglu, E., Deniz, E., James, T., & Kalpathy-Cramer, J., (2010). A novel application of principal surfaces to segmentation in 4D-CT for radiation treatment planning. *Ninth International Conference on Machine Learning & Applications.*

3. Tao, W., & Irene, C., (2009). Fluid vector flow and applications in brain tumor segmentation. *IEEE Transactions On Biomedical Engineering, 56*(3).

4. Rajeev, R., Sanjay, S., & Sharma, S. K., (2009). Brain Tumor detection based on multi-parameter MRI image analysis. *ICGST-GVIP Journal, 9*(III). ISSN 1687-398X.

5. Wang, H., (2009). Three-dimensional medical CT image reconstruction. *International Conference on Measuring Technology and Mechatronics Automation.*

6. Przemyslaw, L., Manuela, P., Mário, M. F., & José, F., (2009). A new 3D image segmentation method for parallel architectures. *IEEE International Conference on Multimedia and Expo.* ICME.

7. Vasupradha, V., Kavitha, A. R., & Roselene R. S., (2016). Automated brain tumor segmentation and detection in MRI using enhanced Darwinian particle swarm optimization (EDPSO). *Procedia Computer Science.* Elsevier.

8. Javeria, A., Muhammad, S., Mussarat, Y., & Steven, L. F., (2017). A distinctive approach in brain tumor detection and classification using MRI. *Pattern Recognition Patterns.* Elsevier.

9. Shrutika, S. H., Akshata, R., & Swati, K., (2017). Implementation of image processing for detection of brain tumors. *International Conference on Intelligent Computing and Control Systems (ICICCS).* IEEE.

10. Manisha, B. R., & Padma, S. L., (2017). Tumor region extraction using edge detection method in brain MRI images. *International Conference on Circuit, Power and Computing Technologies.* IEEE.

11. Annisa, W., Riyanto, S., & Mochamad, M. B., (2018). Brain tumor segmentation to calculate percentage tumor using MRI. *International Electronics Symposium on Knowledge Creation and Intelligent Computing.* IEEE.

12. Atish, C., & Vandana, B., (2018). An efficient method for brain tumor detection and categorization using MRI images by K-means clustering & DWT. *International Journal of Information Technology.* Springer.

13. Mircea, G., Mihaela, L., & Dan, L., (2019). Tumor detection and classification of MRI brain image using different wavelet transforms and support vector machines. *International Conference on Telecommunications and Signal Processing (TSP).* IEEE.

14. Shahriar, S. T. M., Tanzibul, A. K. M., Misbah, U. H., & Mahmuda, R., (2019). Development of automated brain tumor identification using MRI images. *International Conference on Electrical, Computer and Communication Engineering (ECCE).* IEEE.

CHAPTER 9

ADVANCES IN PARALLEL TECHNIQUES FOR HYPERSPECTRAL IMAGE PROCESSING

YAMAN DUA, VINOD KUMAR, and RAVI SHANKAR SINGH

Department of Computer Science and Engineering, IIT (BHU), Varanasi, Uttar Pradesh, India, E-mails: yamandua.rs.cse18@iitbhu.ac.in (Y. Dua), vinod.rs.cse18@iitbhu.ac.in (V. Kumar), ravi.cse@iitbhu.ac.in (R. S. Singh)

ABSTRACT

The recent advancements in the field of computer science that have led to the development of systems with the massive computing power of 1015 floating-point operations per second. High-performance computing (HPC) systems use a large number of computing nodes to attain that performance practically. Parallel computing is a branch of HPC that focuses on reducing the execution time of any application. This work focuses on embedding parallelism in three primary Hyperspectral Image (HSI) processing techniques that are widely used in many applications like object identification, military operations, food security, monitoring natural disasters, and many more. Parallelism in these techniques is required as they work on complex mathematical operations and learning algorithms having high runtime. Some applications of these techniques work on real-time processing having energy and time constraint. It provides a comparative analysis of some recent and trending research works in the field of HSI segmentation, HSI compression, and HSI classification. Important evaluation metrics for parallel algorithms used in subsequent works have been described in detail. The purpose of this chapter is to provide a survey along with future research directions in the field of parallel image processing techniques to get the benefit of parallel computing in HSI processing.

9.1   INTRODUCTION

The recent development in the field of computer science has led to a change in the way computations were performed on standalone systems. Traditional systems were blessed with dedicated resources, and computations were not costly, neglecting the demand for multiple processing units. But the availability of data in large quantities, its generation at a higher speed and development of latest technologies like neural network, deep learning (DL), internet of things (IoT), block-chain, image processing, medical advancements, etc., has increased the demand of fast processing devices. Requirements of these techniques cannot be fulfilled by traditional computation devices like central processing units (CPU) having a single processor. According to Moore's law, the computation power of processors doubles every 2 years. Also, there is a limit on the number of processors that can be used in a circuit due to heat generation and power consumption. These limitations helped the researchers in the last 2 decades to focus more on increasing the speed of transistors by advancement in storage, memory, networking capabilities, and reduced size. But the problem could not be solved until the processors could perform multiple operations at the same time that formed the basis for the development of a different computation system called high performance computing (HPC) that can take the benefit of multiple processors available on board.

9.1.1   HIGH PERFORMANCE COMPUTING (HPC)

High performance computing (HPC) systems are the combination of methods/technologies that enable the use of multiple processors/cores/threads for running advanced programs efficiently. It brings together many concepts like software engineering, electronics, computer architecture, algorithms, and programming libraries to deliver sustained performance using the available resources. An efficient HPC system requires low latency, high bandwidth network for interconnection of multiple nodes and clusters. These systems are constructed by connecting many computers that are either similar in capabilities (called a node) or of different capabilities (called a cluster). Properties of an HPC system include:

- It is a shared resource;
- It can be accessed from a remote location, over a network;
- It has multiple file systems;
- It needs a scheduling mechanism;
- It is a parallel resource;
- It has extensive computation capabilities.

HPC is the use of parallel processing to solve complex computation problems that need a long time or ample resources for execution on a standard computer. It involves the implementation of application programs on Grid computers, Cluster Computers, Distributed computers, multi-core computers, Graphics Processing Units (GPUs), and field programmable gate arrays (FPGAs). It requires two types of modification in sequential programming technique, one at the hardware level, i.e., designing hardware to execute more than one process at a time. The second modification is required at the software level, i.e., reorganization of the program such that it can use the maximum processing power of the hardware. Hardware modifications are already going on, and systems with petaflops or 10^{15} FLOPS are available. The focus of most of the researchers is to develop programs that can efficiently and reliably be executed on these systems and it is only possible by the use of parallel programming. There are many software suites available to help the programer to port the sequential version of a program into a parallel one. But they can only be used after the designing of the parallel algorithm is done and the categorization of parallelism is necessary to understand its details better. There are two broad categories of parallel computation:

1. **Implicit Parallelism:** This is inhibited by processor architecture, compiler, and the operating system (OS). It can be achieved by exploiting the hardware, designing the customized hardware for an application, and optimizing the compiler. It is mainly done by hardware manufacturing companies like NVidia, Intel, etc.
2. **Explicit Parallelism:** It is dependent on the design of the algorithm and method of programming. It is the way in which the algorithm can take benefit of parallel computing.

Both implicit and explicit parallelism can be used in the same application or can be used separately, and then the results can be compared to argue the pros and cons of each. When considering a particular application, it can have multiple levels of parallelism that can be used in parallel computation, like *task-level parallelism, control level parallelism, data-level parallelism, and instruction-level parallelism.* Task level parallelism means dividing a task into multiple subtasks and executing each subtask on a different processor. Control level or function level parallelism states executing multiple functions of the same subtask on different processing units. Data parallelism means dividing data into small independent chunks each processing the same instruction into different units. Instruction level parallelism means executing different operations on the same data chunk on a different thread.

Implementation environment of parallel computing, as considered in this work, can be divided into four categories: shared memory parallelism, distributed memory parallelism, GPU based, and hardware-based parallelism. A system in which multiple processors or computational nodes share the same memory space is called the shared-memory system. In this model, data is shared in a common memory area that is accessible to all the nodes; the problem arises in these systems during simultaneous access to the same memory location. It is parallelized generally by a multi-threading mechanism using some specific library routines (say OpenMP). Distributed memory architecture means every processor or computational node has its own local memory, and data has to be transferred to that memory area before processing. Inter-process communication is managed by passing messages through the connection network. Library routines used for parallelization in these systems are message passing interface (MPI) in C programming language generally categorized as multi-processing model. These systems mainly require proper scheduling mechanism as proposed in Ref. [24] for efficient utilization of resources. GPU is a massive parallelization device that has many cores ranging from a few hundreds to thousands in numbers and privately organized memory. It provides better performance in terms of speed but comes with complications of execution and handling of massive cores without in-depth knowledge of multiprocessor programming. Hybrid systems are the computation networks with all or some of the aforementioned devices having different tools and programming support. Hardware-based parallelism is another aspect of parallel computation that utilizes hardware acceleration for executing parts of an algorithm on specially designed dedicated architecture. FPGAs are such devices that enable parallelism based on dedicated hardware for applications having restrictions over space and resources.

9.1.2 IMAGE PROCESSING

Image processing is the step-by-step procedure to analyze and manipulate the digital image and extract information from it. It plays a vital role in multiple fields of interest like computer science, electronics, optics, medical, mathematics, automobiles, and psychophysics. Its applications in the field of computer vision are remote sensing, clinical image analysis, face detection, biometrics, astronomy, microscope imaging, disaster management, security, meteorology, autonomous driving. Image processing can be categorized into low level, intermediate, and high-level processing. Image operations like histogram analysis, contrast enhancement, filter transformations, noise reduction, contour identification, etc., that extracts description from a digital image is called low-level processing. The output of this processing is also an image and description or contents of the output image are not known in advance. Intermediate level processing comprises of complex operations like object recognition, segmentation, object tracking, region labeling. It converts images into pixel attributes. Results from the intermediate level are taken by high-level processing operations to extract critical information and interpret it in some form. It is called image understanding with multiple applications like pattern recognition, autonomous navigation, object classification, and scene understanding.

Various steps of image processing can be understood from the flow diagram in Figure 9.1, where various steps have their own meaning. There are 11 fundamental steps of digital image processing (DIP) that are described in Figure 9.1.

9.1.3 HYPERSPECTRAL IMAGE

Hyperspectral image (HIS) stores detailed information about the spectral properties of any object or area. These images are a set of contiguous bands, each having a fixed range of wavelength denoting reflectance of the object. Hyperspectral sensors capture light reflected from the object and store the reflectance value band-wise. Each band stores and processes spectral information for some wavelength range, say 400–2,500 nm. Each pixel in the same band corresponds to the same spectral information but varying spatial information, while each pixel at the same spatial position across the band stores reflectance value of the same object for different wavelengths. Spectral information can be the range, resolution of the

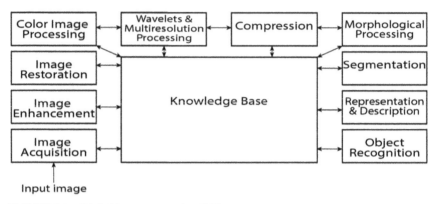

FIGURE 9.1   Digital image processing [25].

spectrum, band adjacency width of each spectral band, or the number of bands. Figure 9.2 represents such an image with multiple objects, each having a set of spectral information associated with it. HSIs are represented using large and multiband data cubes that require large data storage capacities and computational devices for processing and analyzing information. Generally, the size of HSI exceeds hundreds of MegaBytes (MBs) as each pixel stores information of about 12-bit, 16 bit or 32 bits, the number of pixels per band range from a few hundreds to millions, and the number of bands can be in hundreds. For example, calibrated images of standard consultative committee for space data systems (CCSDS) [26] dataset have 224 spectral bands, pixels per band typically count to (677×512), and each pixel stores 16-bit information. The total size for this standard image is ($677 \times 512 \times 224 \times 16$) bits = 148 MB. To identify a pixel, it need not have neighboring pixels; it can be identified by its spectral features alone. The same rule applies to an object, as each object has its own spectral property that differs from others. It helps to reduce the size of data required in object representation. Applications of HSI are in diverse fields; some of them are:

1. **Health Sector:** Disease diagnosis (cancer, retinal, diabetic foot), segmentation of white blood cells (WBC), surgical guidance (visualize surgical bed, monitor oxygen saturation).

2. **Food Quality and Assessment:** Solid contents of blueberries, expired vacuum packaged salmon, oat, and grout kernel, fish evaluation.

3. **Water Resource and Flood Management:** Hydrograph explorations, chlorophyll content, wetland mapping, biochemical contents in water, estimate impact of floods.

4. **Forensic Document Examination:** Ink aging, fraud detection, improve legibility of text.

5. **Artwork Authentication:** Conservation of paintings and its restoration, identification of materials in artwork, identify unique features in art.

6. **Defense and Security:** Detection of the target, distinguish between artificial and natural terrains, detect improvised explosive devices (IED), analyze neurological imbalance.

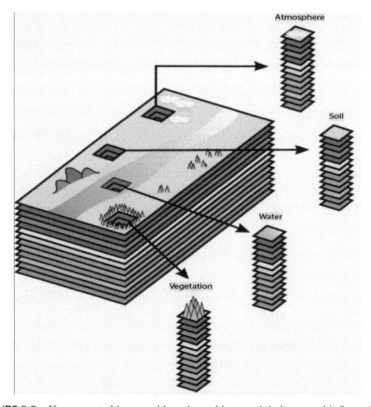

FIGURE 9.2   Hyperspectral image with various objects and their spectral information.

9.1.4 PERFORMANCE METRICS OF PARALLEL COMPUTING

Performance metrics are a set of data and figures used to evaluate the performance of an algorithm. Different metrics are used for different applications, each representing some aspect of the same algorithm. In parallel processing, there are some standard parameters used (along with the application-specific metrics) for the evaluation of parallel algorithms. They are described in subsections.

9.1.4.1 COMPUTATION TIME

It is a measure of the exact time taken by the process in execution. More the computation time, the weaker is the algorithm. It is calculated by calculating the difference of total time taken by a process in execution with the time for which the process was idle (waiting queue, ready queue, I/O queue) in the OS. It can be mathematically represented using Eqn. (1).

$$T_{computation} = T_{execution} - T_{waiting} \qquad (1)$$

9.1.4.2 POWER CONSUMPTION

It is a measure of energy consumed by the hardware device on which the algorithm is being processed. It is calculated by measuring the total power (in watt) taken by the system during the execution of a parallel algorithm.

9.1.4.3 SPEEDUP (THROUGHPUT)

The speedup is a measure of the performance of two systems processing the same task. In parallel computing environment, the speedup is the ratio of computation time taken by a serial algorithm to the time taken by the parallel algorithm for same task maintaining the same volume of the dataset. The speedup is a measure of throughput, i.e., execution rate of the algorithm. The primary objective of the parallel algorithm is to maximize speedup, which is affected by changing the number of processors/nodes. It can be calculated by Eqn. (2). The speedup is directly proportional to the computing performance of the algorithm, i.e., more the speedup, better is the performance.

$$Speedup = \frac{Time\,taken\,by\,best\,sequential\,algorithm(T_s)}{Time\,taken\,by\,parallel\,algorithm(T_p)} \qquad (2)$$

9.1.4.4 EFFICIENCY

It is a measure of the ability of a parallel algorithm to perform proportionally on an increasing number of processors. Efficiency is the ratio of speedup of an algorithm to the number of processors or, in other words, the ratio of time taken by a sequential algorithm to the product of time taken by the parallel algorithm when executed on p processors and number of processors (p). Mathematically it can be represented using Eqn. (3). Its value lies in between 0 and 1 both inclusive. Efficiency equals 1 show the ideal case when speedup is equal to the number of nodes.

$$Efficiency(\eta) = \frac{Time\ taken\ by\ best\ sequential\ algorithm(T_s)}{Time\ taken\ by\ parallel\ algorithm(T_p) * number\ of\ processors} \qquad (3)$$

9.1.4.5 SIZEUP

Sizeup is the measure of the change in the performance of a parallel algorithm on increasing the size of data, keeping the number of nodes constant. It can be calculated using Eqn. (4) and defined as the ratio of the time taken in execution of an algorithm when the original dataset in expanded by p times to the time taken to process the original dataset by the same parallel algorithm [21]. It is used when the size of the dataset is in GigaBytes (GBs); it helps to measure the extent by which algorithm can exploit the performance of a node. More the sizeup of an algorithm, poor is its time effectiveness for big data.

$$Sizeup(p) = \frac{Time\ taken\ by\ parallel\ algorithm\ to\ process\ dataset\ expanded\ by\ 'p'times(T_{pdb})}{Time\ taken\ by\ parallel\ algorithm\ to\ process\ original\ dataset\ (T_{db})} \qquad (4)$$

9.1.5 SCOPE AND CONTRIBUTION

Parallelism has a vast scope in almost all the fields of modern-day applications, but as far as image processing is considered, it plays a vital role in reducing the execution time of image processing algorithms at all stages. Due to multiple mathematical operations involved in it (most of them matrix operations), it can be implemented through specific kernels of GPUs or through hardware accelerators. It also gives a scope to real-time image processing applications (say medical applications) that require an immediate decision on the acquisition site. Big image data like hyperspectral or ultraspectral images can also be processed in real-time with the use of parallel computing.

In this work, significant contributions can be enlisted as:

- A clear concept of parallelism in three primary HSI processing techniques, i.e., image segmentation, image compression, image classification.
- Critical analysis of major recent works in the field.
- A comprehensive overview of HPC, HSI, and evaluation parameters of parallel algorithms.
- Some research challenges and future scope in terms of embedding parallelism in HSI processing techniques.

Next section discusses the parallel image processing techniques in detail with a structured comparison of related researches in image segmentation, HSI compression, and HSI classification techniques.

9.2   PARALLEL IMAGE PROCESSING TECHNIQUES

Image processing techniques can be easily implemented on HPC devices due to the architecture of these algorithms. Most of the image processing techniques work on pixel (point) operations or region-based operations, which can take benefit of multiprocessing behavior of parallel computing. One of the methods to embed parallelism in these algorithms is by dividing the image into multiple parts and executing each part on different processing units (threads, cores, nodes), which can be observed in Figure 9.3. Then combining the results from each unit and merging it as per the demand of algorithm. In this section, the novel methods to parallelize these three essential techniques are studied.

9.2.1   IMAGE SEGMENTATION

Image segmentation can be defined as the process of dividing an image into several groups to unhide the information, like locating an isolated object, interpreting each subgroup separately, and identifying the region of interests (RoI) for future analysis. The process of manual segmentation gives optimal results for small-sized image data but it is not suitable for big data applications like HSI segmentation, medical image segmentation,

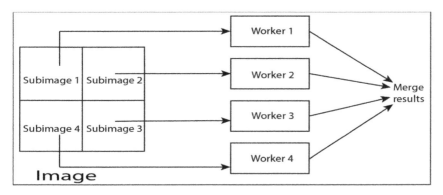

FIGURE 9.3   Parallel image processing.

synthetic aperture radar (SAR) image segmentation, multidimensional classification, and object recognition. These images have a massive number of pixels for which the manual process will be time-consuming and impractical. Optimal image segmentation requires a proper balance of under segmentation and over-segmentation in an application. Image segmentation algorithms can be divided into four major categories: *region-based segmentation, pixel-based segmentation, edge detection based, and hybrid segmentation.* Pixel-based segmentation is most widely used due to its implementational benefits, and it comprises of techniques like thresholding (adaptive, global, etc.), clustering (k-means), and morphology (operations like dilation, erosion, open, close, etc.). Region-based techniques are used to identify regions using algorithms like split and merge watershed, region growing, level set, etc. Edge detection-based segmentation algorithms use curve fitting, and boundary fitting technique to find the drastic change in intensity values of the pixel. The two most essential methods are Gradient-based and Laplacian-based which detect edges by taking the first derivative and second derivative of the image, respectively. HSI segmentation algorithm work on pixel vector that consists of pixel values of all the spectral bands of a spatial position, instead of grayscale value as in the monochromatic image. Examples of segmentation are segmenting alphabets in documents, images of blood cells, celestial bodies in astronomical images, etc. Parallel image segmentation is the implementation of these algorithms in a parallel environment. There has been a lot of research going on in parallel image segmentation; some modern methods are compared in Table 9.1 with each method having its advantages, limitations, methodology, and future directions.

TABLE 9.1 Parallel Techniques for Segmentation of Images

Work Reference	Methodology	Advantages	Limitations	Future Research Directions
Sui-Hua et al. [1]	• The scattering matrix is used to calculate the scattering properties of the image • 2D Convolution Neural Network (CNN) with a filter of 3 × 3 size is applied to the image • The proposed algorithm is then implemented on GPU	• Implementation on GPU provides an average speed-up of 176 times • Better accuracy compared to state-of-the-art techniques	• Post-processing of results is not considered • Fixed patch size of 21 × 21	• Use stochastic pooling technique to have better performance • Use the multipath CNN to improve the accuracy further
Fang et al. [2]	• Convert image into LUV color space • Select a sample point and a kernel function to calculate its vector value • If the convergence at that point is considerable select another sample point and repeat the procedure until all points are covered • Else change the kernel and repeat the procedure • MPI and OpenCL are used to schedule and calculate the segmented region of all points on GPU and GPU cluster	• Stable speedup on single GPU as well as on GPU cluster • Scalable to change detection algorithms in multi-temporal images	• Works only for a series of remote sensing images, not for a single image	• Use of dynamic load-balancing algorithm to schedule the task among GPU and CPU
Hossam et al. [3]	• Two different approaches are used to segment the image on GPU • The first approach utilizes one thread of GPU to calculate all dissimilarities of a region • The second approach uses one thread to calculate dissimilarity between only one pair of regions	• Reduced energy consumption and efficient implementation on cluster	• Dissimilarity calculation is the most complex task	• Use of dynamic programming to avoid re-computation • Exploit vectorization and loop unrolling to reduce time further

TABLE 9.1 *(Continued)*

Work Reference	Methodology	Advantages	Limitations	Future Research Directions
Rahul et al. [4]	• Number of clusters is identified by histogram analysis using the Otsu method • 10 stage Moore machine is used to apply the k-means algorithm • Verify the segmentation by comparing the results with true land features	• Reduced hardware requirements even for less number of clusters • Proposed hardware consumes less power and requires less area	• Segments the images only based on its chromatic property	• Can be applied to hyperspectral images by considering the spectral domain
Haiyan et al. [5]	• Form graph considering each pixel as a vertex and each edge as a pair of neighboring pixels • Obtain minimum spanning tree from it based on some threshold function • Merge adjacent regions using minimum heterogeneity rule algorithm using some heterogeneity function • MPI based data parallelism is used to reduce the execution time of the algorithm	• Suitable for multiple landscapes • Shape heterogeneity is considered along with color • 7.65 speedup is achieved by 14 processing units	• Parameters are selected by a hit-and-trial method • Efficiency decreases by increasing the number of processes	• Development of automation method for parameter selection • Address boundary problem of parallel segmentation
Biao et al. [6]	• Divide the image into multiple sub-images of the same size and process each subdivision on different processing units • Extract superpixels within each subdivision and apply graph-based segmentation considering each superpixel as a node • Finally, merge the regions with similar properties	• Reduces the effect of speckle-noise by considering global information • Reduced calculations by considering each superpixel as a node	• Memory requirements are not considered • Identification of superpixel is also restricted to the subdivision in which it falls	• Implementation on GPU by considering task parallelism

9.2.2 IMAGE COMPRESSION

Image compression is the technique of reducing the size of the digital image by storing the information in some other form by using less number of bits than original data. Traditional image compression algorithms take an image as input and produce encoded bitstream as output. The bitstream can be transmitted/stored using less bandwidth/memory, and the original image can be reconstructed at any point in time. These algorithms can be classified into lossy or lossless based on the quality of the desired reconstruction. If the original image is required without any loss of information, the algorithm used is called lossless algorithms. It is mainly used for specific applications where any loss to data is not tolerable, due to this reason only a small compression performance can be achieved. When some amount of information loss is acceptable, lossy algorithms can provide better performance by providing the original image with some information loss and degradation after reconstruction. Digital grayscale or RGB image use these traditional techniques.

HSI compression is a technique through which the size of HSI can be reduced without loss in image quality beyond the desired level [23]. It is one of the essential steps of HSI processing, which is included in every application. It reduces the cost of bandwidth and storage equipment. Compression reduces the size by storing the same information with a small number of bits. It uses different representations and removes redundancy existing in the image. High redundancy helps compression algorithms to achieve a high compression ratio. There are three types of redundancies exists in an HSI:

1. **Spatial Redundancy:** It arises due to intra-band dependency that exists in spatial domain;
2. **Spectral Redundancy:** It arises due to dependency among pixels of different band at same spatial location; and
3. **Temporal Redundancy:** It arises when HSI of the same location are taken at different time, dependency in temporal domain (for same spectral and spatial pixels) results in temporal redundancy.

These redundancies are decorrelated in compression algorithms, and thus data size is reduced. Original data can be reconstructed using decompression which is usually the reverse process of compression. HSI compression techniques can be classified into the following major categories:

transform-based compression, compressive sensing-based compression, tensor-decomposition based compression, vector-quantization based compression, and prediction-based compression techniques that exploit spectral correlation existing in HSI along with the spatial correlation. Transform based technique is the most popular 2D image compression technique that has been extended to 3D or HSI compression. It is known as a transform-based technique as it transforms the pixels values into the frequency domain by applying some transformation function to all three dimensions of HSI. Vector Quantization is a data compression technique that takes 3D HSI data cube as input and returns a compressed image. Two significant steps of this method are training (codebook generation) and coding (code vector matching). It represents the pixel values of any spatial position of the first band as the head of an n-length vector that consists of pixels of n-different bands, where n=total number of bands of HSI. It quantifies the vector instead of performing decorrelation. The tensor decomposition technique is one of the latest techniques for image compression, which gives high performance compared to traditional methods. Tensor could be considered as an n-Dimensional matrix that can be decomposed easily. In this technique, HSI is stored into 3D tensor(Y), and one of the tucker decomposition techniques is applied to decompose the 3D tensor(Y) into lower dimension 3D tensor(X). Decomposed tensor is then encoded and transmitted through the channel. The compressive sensing technique is famous for onboard compression algorithms as it shifts the computation time of encoder to the decoder. It is used in real-time compression as it senses a small chunk of data, compresses it, transmits it to the receiver, and then accepts another chunk. Prediction based compression is an alternative to transform based algorithms with technical and implementational benefits. In this technique, the value of a pixel is predicted after applying some mathematical functions to the previous pixel values. It is developed especially for 3D images, exploits correlation in both spatial and spectral directions, and removes them. Prediction in HSIs is mainly applied on spectral-domain with the help of a filter after spatial decorrelation gets completed. The time complexity of all these algorithms is very high; thus, the main focus of parallel HSI compression algorithms is to reduce the execution time. Table 9.2 consists of the recent work in the field of parallel HSI compression with a critical analysis of all considered algorithms.

Table 9.2. Parallel Techniques for Compression of HSI

Work Reference	Methodology	Advantages	Limitations	Future Research Directions
Maria et al. [7]	• Parallel implementation of HyperLCA algorithm • Comprises of three main steps: copy of data from CPU to GPU memory, launch of seven different kernels to perform the seven steps of transform, and at last copy of transformed results back to CPU memory • Transform, and decoding has been performed independently on two different CPU processes	• Implementation on low power graphical processing units (LPGPUs) • Compresses each frame in real-time, i.e., compression time is less than capture time	• CPU and GPU share the same physical RAM • Works for sensors that generate at most 256 bands HSI	• Can be implemented to compress medical HSI
Alfonso et al. [8]	• HSI dataset is partitioned into segments, and each segment is compressed separately in parallel • Two different methods of partitioning are applied to achieve data parallelism: strip based partitioning and square-based partitioning • Strip based partitioning is applied over y-axes that produce non-uniform sub-images while square-based partitioning produces uniform sub-images	• A scalable solution, i.e., efficiency, and throughput can be increased by increasing the number of hardware • Exploits data-level parallelism	• Dependent on device architecture • Introduces energy efficiency and speedup as a tradeoff factor	• Follow a similar approach for hardware implementation of other state-of-the-art compression algorithms
Jiaojiao et al. [9]	• Form cluster of spectral lines into M classes • Calculates prediction coefficients and then prediction image from it • Encode the residual image and prediction coefficients • Executed complex matrix multiplication, inverse, determinant on GPU	• Reduces complexity by parallel processing • Three different techniques of parallel implementation proposed	• Complex implementation	• Implementation of the algorithm on FPGA following the same fashion • Optimization of parameters

TABLE 9.2 (Continued)

Work Reference	Methodology	Advantages	Limitations	Future Research Directions
Ali et al. [10]	• Consists of intraband encoding, superpixel segmentation • Followed by vectorization, Recursive Least Square (RLS) prediction, and entropy encoding • Implementation of the algorithm in parallel by executing each superpixel on different units	• Parallel implementation with 12 parallel workers and changing vector length	• ROI selection is a manual and complicated task	• Implementation on GPU and its critical evaluation
Wenbin et al. [11]	• Group the spectral bands into $N/(n+2)$ groups, where N is the total number of bands in HSI and n is the number of processing units • Allot one group to each of the processing units to calculate the prediction coefficients using the 2D mean square in parallel	• Makes full use of system resources • Reduced execution time by 3.92 times using 4 processors	• Spatial correlation not considered • Intergroup correlations not considered	• Can improve compression ratio by considering the correlation in the spatial domain
Wenbin et al. [12]	• Uses k-means clustering algorithm to group hyperspectral image along a spatial dimension and convert them into 2-D matrices • Then use the adaptive method, i.e., use two previous bands to predict the current pixel value if the spectral coefficient > 0.9, else original data is kept directly	• Compression performance is dependent on the number of clusters, which can be increased.	• Time taken to cluster the image is not taken into account • Manual selection of the number of clusters	• Can improve the speedup by utilizing more available processors by parallel implementation of the k-means algorithm
Marius et al. [13]	• Divide the original HSI along column axes and spectral domain • Apply prediction algorithm to each partition in parallel but execute neighboring blocks of column axes in first in first out (FIFO) order • Again, divide predicted coefficients along the spectral domain and apply Golomb coding in parallel	• An efficient way of parallel prediction along with parallel encoding is proposed • Utilizes data parallelism to decrease the total execution time of the algorithm	• Data dependency at the prediction stage hinders the full utilization of resources	• Improve the memory efficiency by considering caching strategy

9.2.3   IMAGE CLASSIFICATION

Classification is an essential stage of image processing that helps to categorize an image or part of it on the basis of some a priori knowledge. It is used in many applications like facial recognition, disease identification, image search engines, theft identification, etc. It is an essential method of information retrieval with the objective of classifying the pixels into some category. HSI classification algorithms can be categorized as *supervised and unsupervised, fuzzy, and crisp, parameterized, and non-parameterized classification*. Supervised classification can be understood as the method in which images are classified based on some apriori knowledge (ground truth) to train, test, and validate the model. Minimum distance, fisher classification, maximum likelihood discrimination, etc., are some of its techniques. Unsupervised classification deals with the pixel values only on the basis of characteristics of data and classifies them into different classes. The process of assigning a pixel of an HSI to only one specific category is called hard or crisp classification. Traditional methods were based on this method. When one pixel is assigned to more than one class at a time, the method is called fuzzy classification. This method is mainly used in the clustering of images based on the fact that HSI can have fuzziness. The process follows a series of steps beginning from randomly initializing the center of each cluster and then iteratively approximating the right center by finding the difference between degrees of a pixel in each category with the previous iteration until the degree is smaller than a threshold value. Parameterized classification is a method based on the probability distribution function (PDF) of each class of HSI. The maximum likelihood classification and nearest mean value are an example of this category that estimates the distribution parameter of each class and classifies an image based on it. Non-parameterized classifiers differ in the way that they do not need the assumption of any category or class. Parallelepiped classification and neural network classifier belong to this category. Parallel HSI classification generally reduces the performance of sequential algorithms as each node might not have a complete set of bands and a complete set of neighboring pixels at the boundary of the division of the dataset. So, they need additional care in this part and should focus more on increasing the classification performance. Some crucial developments in parallel HSI classification are compared in Table 9.3, each with future research directions.

TABLE 9.3 Parallel Techniques for Classification of HSI

Work Reference	Methodology	Advantages	Limitations	Future Research Directions
Pedro et al. [14]	• Implementation of the preprocessing stage of classification on CPU and GPU environment using OpenMP and CUDA library • Different built-in kernels are used for attribute opening profile • The wavelet-based feature extraction method is used to reduce the dimension	• Reduced memory requirements using matrix-based data structure • The small profile also keeps most spectral information	• Value of pixel is scaled between 0 and 255 losing minute details • Steps of CPU and GPU are the same	• Built-in kernels can take advantage of hardware accelerators and thus can be used for on-site classification
Lei et al. [15]	• The local adaptive weighted average value is replaced with the central pixel of a window • Coefficients are then estimated by sparse representation model • To consider and protect spectral information, it uses L_1 minimization as spectral consistency constraint	• Preprocessing creates more discriminative data, i.e., reduced noise and spectral variations • Reduced training dataset	• Many steps are executed on the CPU, that takes most of the time	• Reduce the time taken in data transfer from host to device • Increase the training size to see if it could further improve accuracy
Haicheng et al. [16]	• 3D convolution and max pool layer generate a 1-D vector of spectral-spatial features which are assembled into 2D feature map • 2D CNN is then applied followed by fully connected, dropout, and SoftMax layer • The entire process is parallelized by the master-slave cluster in which parameters and data is shared to reduce the execution time	• Combination of 2-D and 3-D CNN gives excellent performance accuracy • Each worker updates parameters to master so reduced communication during execution	• Unstable speedup for implementation on more than 3 GPUs	• Implement the algorithm on a large number of workers and reduce communication overhead

TABLE 9.3 *(Continued)*

Work Reference	Methodology	Advantages	Limitations	Future Research Directions
Emanuele et al. [17]	• Two different approaches have been proposed for the training of autoencoders • One with OpenMP API using the thread-based processing to reduce the execution time by working on the same sized data chunk • Second with CUDA API, that uses *cublasDgemm* function for matrix multiplication on GPU	• Implementation on two different families of GPUs, i.e., Kepler, and Volta • Dependent on hardware architecture, thus scalable	• Maximum speed up using OpenMP is 4 • Speedup of GPU is also dependent on training size	• Implementation of other classification techniques in parallel using similar methodology
Shuanglong et al. [18]	• Convolution layer and fully connected layers have been implemented on hardware units • Each layer is processed only when off-chip transfers the input and weight of each layer to on-chip memory	• Significant speedup achieving real-time speed • Improved accuracy compared to other FPGA designs	• Performs pixel–wise processing on–chip buffers	• Hardware accelerators for depth wise convolution layer
Beatriz et al. [19]	• Transform the HSI into hue saturation value (HSV) and Normalized Difference Vegetation Index (NDVI). • Extract feature image with 11 features • Apply ANN on two GPU cluster executing HSV and NDVI image in parallel	• Time taken to transfer data between host and device is negligible in 2-GPU cluster	• Hyperspectral signal is separated into visible and near-infrared range	• Optimize the parameters of ANN
Raquel et al. [20]	• SS classification algorithm is used to classify medical HSI • 1st step of the algorithm is principal component analysis (PCA) to reduce the spectral dimensions • 2nd step is to use support vector machine (SVM) for classification, and	• Two different applications with dissimilar requirements: one with the time constraint, other with energy constraint	• Memory requirement is not considered • No performance improvement on	• Apply more promising algorithms like multidimensional CNN to improve the classification accuracy

TABLE 9.3 *(Continued)*

Work Reference	Methodology	Advantages	Limitations	Future Research Directions
	• In the 3rd step, k-nearest neighbor (KNN) is applied for spatial filtering of the results • The algorithm is then executed on three different platforms using a different methodology	• Uses three different families of parallel processing devices	increasing data size for low-power platforms	
Ji'an et al. [21]	• Spark-based parallel computing was used for parallel execution of ANN and SVM classification algorithm • The cluster has six servers; one master and 5 slaves were used after the preprocessing stage	• Improved operational speed of algorithms by using Scala, and spark framework • Adaptive solution, i.e., use SVM for binary classification and ANN for multi-class classification	• Manual selection of ROI • Library dependent implementation	• Use of complex ANN to improve the accuracy • Also, the speedup can be improved by considering more number of nodes
Kun et al. [22]	• Gaussian–Bernoulli restricted Boltzmann machine (GBRBM) uses an unsupervised method to extract features from spectral-domain of HSI • Multiple GBRBM models are applied to the HSI with different hidden neurons on multiple processors in parallel • Output features are combined and given for logistic regression classification, where classes are labeled	• Increase in accuracy by 2–5% (96.22% in Pavia university dataset) • Short prediction time	• Longer training time despite reduced training data	• Use of spectral-spatial feature fusion to improve the accuracy

9.3   DISCUSSION AND OPEN CHALLENGES

HSI processing techniques considered in this work are the most important ones, and they form the base for many applications. They cover a broad range of problems from acquisition at source to identifying the objects, from the prediction of diseases in medical applications to observing the quality of food. As discussed in Section 9.1, the HPC implementation environment can be categorized into four broad ways, shared memory, distributed memory, GPUs, and hardware-based accelerator. The above mentioned three broad techniques and the articles considered can be further categorized based on the implementation environment. Table 9.4 describes the significant contribution in each field according to the parallel environment in which they have been implemented. This study gives details of the fact that the field of parallel HSI segmentation is still in a very nascent phase, and has more gap compared to compression and classification. As the focus of researchers has been to SAR image, HSI segmentation can take benefit of existing algorithms mainly in the field of shared memory parallelism. The design of FPGA can also be explicitly optimized to segmentation algorithms to obtain better speedups with reduced power consumption. HSI compression algorithms are mainly required onsite, i.e., at the site of acquisition especially in remote sensing applications where both the memory constraint and power constraint need to be tackled. In medical images, the problem is to maintain the quality of the reconstructed image, specifically of some critical regions. The region-based compression techniques can be optimized for hardware and multithreading parallelism. The researchers have found optimal results in the classification subdomain

TABLE 9.4   Image Processing Techniques Implemented in Different Environment

Implementation environment \ Techniques	Segmentation	Compression	Classification
Shared memory	[6]	[10], [13]	[14]
Distributed memory	[2], [5]	[12], [11]	[16], [19]
GPU	[1], [2], [3]	[9], [7]	[14], [15], [16], [17], [19], [21], [22]
FPGA (hardware based)	[4]	[8]	[18], [20]

for GPU parallelism, but the issue is to maintain the performance in terms of overall accuracy, kappa coefficient, etc. Moreover, some object identification techniques at the sensor level have been proposed that necessarily need to be operated in real-time, with minimum use of computation resources.

9.4 CONCLUSION

HPC is needed in almost all the stages of image processing techniques starting from data acquisition through sensors to image representation and description. The scope of this work is limited to only three stages namely image segmentation, image compression and image classification and their parallelization through parallel computing. These techniques have been selected as they form the base in many applications and must be completed in short time duration for real-time systems. In this work, the significant researches in these three categories are analyzed, identifying the future scope in all the considered articles. The focus has been mainly on HSIs due to their large size and application benefits. The algorithms have also been categorized based on their implementation environment to provide a comparative study of the implementation environment. Algorithms of different techniques are listed with their methodology and limitations compactly. Such classifications could help in the development of advanced compression algorithms and may boost many programs related to clinical applications, food security, remote sensing space operations, and many more.

KEYWORDS

- **central processing units**
- **field programmable gate arrays**
- **graphics processing units**
- **high performance computing**
- **internet of things**
- **megabytes**

REFERENCES

1. Wang, S. H., Sun, J., Phillips, P., Zhao, G., & Zhang, Y. D., (2018). Polarimetric synthetic aperture radar image segmentation by convolutional neural network using graphical processing units. *Journal of Real-Time Image Processing, 15*(3), 631–642.

2. Huang, F., Chen, Y., Li, L., Zhou, J., Tao, J., Tan, X., & Fan, G., (2019). Implementation of the parallel mean shift-based image segmentation algorithm on a GPU cluster. *International Journal of Digital Earth, 12*(3), 328–353.

3. Hossam, M. A., Ebied, H. M., Abdel-Aziz, M. H., & Tolba, M. F., (2018). Accelerated hyperspectral image recursive hierarchical segmentation using GPUs, multicore CPUs, and hybrid CPU/GPU cluster. *Journal of Real-Time Image Processing, 14*(2), 413–432.

4. Ratnakumar, R., & Nanda, S. J., (2019). A low complexity hardware architecture of K-means algorithm for real-time satellite image segmentation. *Multimedia Tools and Applications, 78*(9), 11949–11981.

5. Gu, H., Han, Y., Yang, Y., Li, H., Liu, Z., Soergel, U., & Cui, S., (2018). An efficient parallel multi-scale segmentation method for remote sensing imagery. *Remote Sensing, 10*(4), 590.

6. Hou, B., Zhang, X., Gong, D., Wang, S., Zhang, X., & Jiao, L., (2017). Fast graph-based SAR image segmentation via simple super pixels. In: *2017 IEEE International Geoscience and Remote Sensing Symposium* (pp. 799–802). IEEE.

7. Díaz, M., Guerra, R., Horstrand, P., Martel, E., López, S., López, J. F., & Sarmiento, R., (2019). Real-time hyperspectral image compression onto embedded GPUs. *IEEE Journal of Selected Topics in Applied Earth Observations and Remote Sensing, 12*(8), 2792–2809.

8. Rodriguez, A., Santos, L., Sarmiento, R., & De La Torre, E., (2019). Scalable hardware-based on-board processing for run-time adaptive lossless hyperspectral compression. *IEEE Access, 7*, 10644–10652.

9. Li, J., Wu, J., & Jeon, G., (2019). GPU acceleration of clustered DPCM for lossless compression of hyperspectral images. *IEEE Transactions on Industrial Informatics*.

10. Karaca, A. C., & Güllü, M. K., (2019). Superpixel based recursive least-squares method for lossless compression of hyperspectral images. *Multidimensional Systems and Signal Processing, 30*(2), 903–919.

11. Wu, W., Wu, Y., & Qiao, X., (2018). Parallel compression based on prediction algorithm of hyper-spectral imagery. In: *MATEC Web of Conferences* (Vol. 173, p. 03070). EDP Sciences.

12. Wenbin, W., Wu, Y., & Li, J., (2018). The hyper-spectral image compression based on k-means clustering and parallel prediction algorithm. In: *MATEC Web of Conferences* (Vol. 173, p. 03071). EDP Sciences.

13. LOlaru, M., & Craus, M., (2017). Lossless multispectral and hyperspectral image compression on multicore systems. In *2017 21st International Conference on System Theory, Control and Computing* (pp. 175–179). IEEE.

14. Bascoy, P. G., Quesada-Barriuso, P., Heras, D. B., Argüello, F., Demir, B., & Bruzzone, L., (2019). Extended attribute profiles on GPU applied to hyperspectral image classification. *The Journal of Supercomputing, 75*(3), 1565–1579.

15. Pan, L., Li, H. C., Ni, J., Chen, C., Chen, X. D., & Du, Q., (2018). GPU-based fast hyperspectral image classification using joint sparse representation with spectral consistency constraint. *Journal of Real-Time Image Processing, 15*(3), 463–475.

16. Qu, H., Yin, X., Liang, X., & Liu, W., (2018). Parallel dimensionality-varied convolutional neural network for hyperspectral image classification. In: *International Conference on Intelligence Science* (pp. 302–309). Springer, Cham.

17. Torti, E., Fontanella, A., Plaza, A., Plaza, J., & Leporati, F., (2018). Hyperspectral image classification using parallel autoencoding diabolo networks on multi-core and many-core architectures. *Electronics, 7*(12), 411.

18. Liu, S., Chu, R. S., Wang, X., & Luk, W., (2019). Optimizing CNN-based hyperspectral image classification on FPGAs. In: *International Symposium on Applied Reconfigurable Computing* (pp. 17–31). Springer, Cham.

19. Garcia-Salgado, B. P., Ponomaryov, V. I., Sadovnychiy, S., & Robles-Gonzalez, M., (2018). Parallel supervised land-cover classification system for hyperspectral and multispectral images. *Journal of Real-Time Image Processing, 15*(3), 687–704.

20. Lazcano, R., Madroñal, D., Florimbi, G., Sancho, J., Sanchez, S., Leon, R., & Marrero-Martin, M., (2019). Parallel implementations assessment of a spatial-spectral classifier for hyperspectral clinical applications. *IEEE Access, 7*, 152316–152333.

21. Xia, J. A., Yang, Y., Cao, H., Zhang, W., Xu, L., Wang, Q., & Huang, B., (2018). Hyperspectral identification and classification of oilseed rape waterlogging stress levels using parallel computing. *IEEE Access, 6*, 57663–57675.

22. Tan, K., Wu, F., Du, Q., Du, P., & Chen, Y., (2019). A parallel gaussian–Bernoulli restricted boltzmann machine for mining area classification with hyperspectral imagery. *IEEE Journal of Selected Topics in Applied Earth Observations and Remote Sensing, 12*(2), 627–636.

23. Principles of Remote Sensing-Center for Remote Imaging, Sensing and Processing, Crisp. *Analog and Digital Images*. https://crisp.nus.edu.sg/~research/tutorial/image.htm (accessed on 30 September 2021).

24. Gupta, S., Agarwal, I., & Singh, R. S. (2019). Workflow scheduling using Jaya algorithm in cloud. *Concurrency and Computation: Practice and Experience*, e5251.

25. Gonzales, R. C., & Woods, R. E., (2002). *Digital Image Processing*. (2nd ed.). Addison-Wesley Pub (Sd).

26. NASA. *123.0-B-Info TestData*. URL: https://cwe.ccsds.org/sls/docs/sls-dc/123.0-B-Info/TestData (accessed on 30 September 2021).

CHAPTER 10

CASE STUDY: PULMONARY NODULE DETECTION USING IMAGE PROCESSING AND STATISTICAL NETWORKS

JIGNYASA SANGHAVI

Department of Computer Science and Engineering,
Shri Ramdeobaba College of Engineering and Management, Nagpur,
Maharashtra, India, E-mail: sanghavijb1@rknec.edu

ABSTRACT

According to World Health Organization, cancer is the second leading cause of death globally, and is responsible for an estimated 9.6 million deaths in 2018. Globally, about 1 in 6 deaths is due to cancer. Among cancers, lung cancer is the leading cause of death. Lung cancer is the abnormal and uncontrolled growth of cancerous cells in the lungs. These cells destroy the nearby tissues, and cells form cancerous nodules. Automatic Computerized detection systems are used for the early detection of Lung cancer. In this chapter, the systematic review of automatic detection of the pulmonary nodule from Chest CT scans using image processing and machine learning techniques is illustrated. The automatic CAD (computer-aided diagnosis) systems from simple CBIR systems using similarity functions to complex deep neural networks with false-positive reduction techniques are summarized in this chapter. This chapter gives a brief introduction of methods and techniques used in the processing stages of an automatic system like pre-processing, segmentation, feature extraction, feature optimization, and classification.

10.1 INTRODUCTION

Lung cancer is the leading cause of death among cancers because it does not show any sign or symptoms in early stage. Abnormal growth of tumor cells leads to development of nodules in lungs. The size of these nodules varies from 3 mm to 30 mm. The imaging modalities used for pulmonary nodules detection are CT scans, PET or MRI. However, CT scan shows the clear imaging of organs, tissues in lung, cancerous parts clearly and is less expensive as compared to MRI and PET so generally CT scans are used for early detection. CT scans are divided into slices and radiologist needs to read it slice wise. Figure 10.1 shows the CT scan image of lungs with pulmonary nodule.

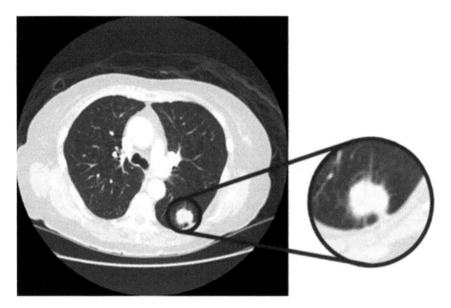

FIGURE 10.1   Pulmonary nodule in CT scan of lungs.
Source: Image courtesy by Gupta et al. [11].

Automatic computer-aided diagnosis (CAD) systems can be used by health professionals as second opinion. For developing Automatic CAD systems various image processing and statistical machine learning (ML) algorithms, optimization algorithms and deep neural networks (DNNs) are used. The basic workflow mechanism of automatic CAD system is depicted in Figure 10.2 showing all the phases of the system.

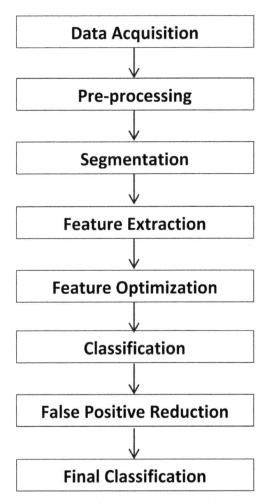

FIGURE 10.2   Basic workflow mechanism of automatic CAD systems.

Recently deep CNN have achieved the great success as it performs all the stages in one or combination of networks. Handcrafted feature extraction or segmentation is not required in DNNs. DNNs are special type of artificial neural networks (ANNs) which takes input as image, assigns weight and bias according to image and able to classify the image. Very limited pre-processing of image is required and remaining all the phases is taken care by CNN. CNN is shift invariant and space invariant ANNs. The computational complexity of CNN is very large because it has many layers and computation starts with whole image matrix and then concluding on classification. Figure 10.3 shows working of CNN.

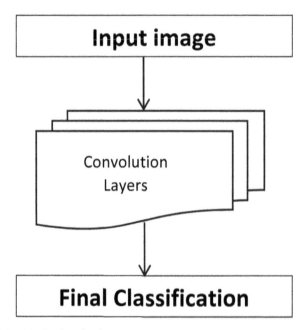

FIGURE 10.3   Mechanism for deep neural networks.

10.2   LITERATURE REVIEW

Nodule detection is basically categorized in different perspectives. Classification with respect to (i) size is from micro nodules to mass nodules, (ii) types like solid, non-solid, and sub-solid and (iii) location like Juxta Vascular to Juxta Pleural. This section summarizes the different methods of automatic pulmonary nodule detection using image processing, ML techniques and deep learning (DL) techniques. The automatic detection is basically performed with data acquisition, pre-processing, segmentation, feature extraction and classification.

10.2.1   DATA ACQUISITION

Among all the medical imaging modalities CT is preferred for diagnosis of lung cancer. Some of the researchers have used dataset taken from hospitals such as Walter Cantídio University Hospital, Fortaleza, Brazil containing 36 Chest CT images [3]. To validate the performance of the proposed system with respect to standardized format mostly the researchers have taken available datasets for their research. The commonly used datasets by researchers are: (i) LIDC-IDRI (lung image database consortium and image database resource initiative) dataset from The Cancer Imaging Archive (TCIA) composed with 1,010 patients, 1,018 CT exams and 244,527 images with lung nodules obtained from seven institutions [1, 2, 4, 7, 9, 10]. (ii) The NLST [12] dataset contains scans from the CT arm with low-dose CT [8]. (iii) LUNA16 is classified dataset used for supervised learning and it is subset of LIDC-IDRI dataset having 754,976 candidates [1, 7]. (iv) VIA/ELCAP (Vision and Image Analysis group and Early Lung Cancer Action Program) dataset containing 50 low dose whole lung CT scans [6]. (v) Cancer imaging archive (CIA) dataset contains the images contributed by the national cancer institute tumor analysis consortium lung cohort with 5,043 images [5].

10.2.2   PRE-PROCESSING OF CT IMAGES

Medical imaging modalities are generated using X-rays or other rays and therefore preprocessing is required either before segmentation or after segmentation. However, publicly available databases are standardized datasets which can be directly given as input to deep networks. Some of the researchers have used preprocessing for noise removal and image enhancement. Moghaddam et al. [2] has used the Gaussian kernel for filtering then by BSE (blob structure enhancement) filter to extract initial candidates on segmented image. Line tracking method, followed with LSE (line structure enhancement) and CAM (central adaptive medialness) filter is used for preprocessing the segmented image. Shakeel et al. [5] have used weighted mean histogram equalization approach for noise removal and histogram equalization for image enhancement. Lakshmanaprabhu et al. [6] have used median filtering and adaptive histogram equalization for noise removal and image enhancement in pre-processing phase.

10.2.3 SEGMENTATION FROM CT IMAGES

Chest CT scan consist of blood vessels, respiratory vessels, arteries, left, and right lung, therefore segmentation of region of interest (RoI) becomes necessary if not using deep networks. Generally thresholding is used for making filters of lung and then image is segmented using that filter. In proposed system [2, 9], first segmentation is performed on image to get the RoI using thresholding with two masks and then preprocessing is applied. In the system [5], after image enhancement, the RoI is segmented by applying the improved profuse clustering technique (IPCT) based on the Spectral super pixel clustering technique.

10.2.4 FEATURE EXTRACTION

Features describe the characteristics of the RoI in terms of its color, shape, and texture. Feature defines the behavior of image and decides the efficiency of the classification. After segmentation, the features are extracted from the segmented RoI for providing input to classifier. Different combinations of shape, region, and texture descriptors are extracted for giving input to classifiers other than deep networks. Out of 10 papers selected some of the researchers extracted features from geometric features, intensity features, texture descriptor, gradient feature and region descriptor [2, 9], Gabor features (High contrast features), Zernike (polynomial decompositions) and statistical based texture features [3], 3D texture attribute, Margin sharpness attribute and integrated of both attributes [4], spectral features like mean, standard deviation, Third Moment skewness, Fourth Moment Kurtosis [5], histogram features like Variance, mean, skewness, and kurtosis, standard deviation, texture features like GLCM and wavelet features [6].

10.2.5 FEATURE OPTIMIZATION

This phase is optional; some researchers have implemented feature selection or reduction techniques to reduce the computational complexity. In this, high dimensional feature vectors extracted are reduced and selected features are given input to classifier. Some of the feature selection techniques used by researchers are: sparse coding method for feature learning which is unsupervised learning used in DL [2], bio-inspired algorithms for feature selection like Improvised Crow Search Algorithm, Improvised Cuttlefish Algorithm, and Improvised Grey Wolf algorithm [3], Z-normalization [4], linear discriminant analysis (LDA) [6], and cross correlation analysis [9].

10.2.6 CLASSIFICATION

Classification is the final process to classify whether the nodule is cancerous or non-cancerous. Some of them have used similarity analysis by simply using distance function in CBIR (content-based image retrieval) system [4]. Some of the advanced classifiers used by researchers for high dimensional feature set are linear regression model [2], ML classifiers like K-nearest neighbor (KNN), support vector machine (SVM), random forest (RF), and decision tree (DT) [3], genetic algorithm (GA) in combination with particle swarm optimization (PSO) [9].

10.2.7 DEEP NEURAL NETWORK (DNN)

A DNN is a complex neural network with high level of complexity and containing more than two layers. It is designed such that when image is given as input all processing phase including segmentation, feature extraction and classification all are performed with one network. Recently DNNs are used in many classification domains. Some of the researchers have used DNNs for all the processing and some have used different DNNs for each processing phase. In proposed system [1], three 2D convolution networks are implemented on different CT slices so as to have 3D view of the nodules. This increases the sensitivity for detecting the individual nodules in the slice. Improved Faster R-CNN is used which is further integrated with deconvolution layer for getting enlarged feature map. DL instantaneously trained neural network (DITNN) used for training the images with high dimensional features [5]. Lakshmanaprabhu et al. [6] implemented optimal deep neural network (ODNN) designed using deep belief network (DBN) and restricted Boltzmann machine (RBM). For weight optimization and to reduce the hidden layers modified gravitational search algorithm (MGSA) is used with ODNN. Some of the researchers have used standard architectures like the U-Net architecture for complete process of segmentation, feature extraction and classification [7]. Marysia Winkels et al. [8] has used baseline convolution neural network (CNN) with Group Convolution network for complete classification process. Yutong et al. [10] propose a semi-supervised adversarial classification (SSAC) model which consists of an adversarial autoencoder-based unsupervised reconstruction network R, a supervised classification network C, and learnable transition layers that enable the adaption of the image representation ability learned by R to C.

10.2.8   FALSE POSITIVE REDUCTION

False positive are nodules detected as cancerous but are non-cancerous. In medical imaging, it is very risky if there is false diagnosis. Some of the researchers have implemented statistical methods for false positive reduction. In the system implemented [1] for false positive reduction stage, Boosting was combined with CNNs. In this design they have used three 2D CNNs for three different slices of CT scan. Therefore, the final result is obtained by voting. Boosting is a commonly used statistical learning method which is effective and also has a broad application.

In the proposed system [8] 3D group equivariant convolutional neural networks (G-CNNs) is used for false positive reduction along with baseline CNN for lung nodule detection.

10.3   CONCLUSION

In this chapter, we tried to present a novel and effective computer-aided automated pulmonary nodule detection systems and CBIR systems for pulmonary nodules. The working methods based on similarity matching to complex CNN were summarized in this chapter. The statistical methods used in each stage of automatic detection like pre-processing, segmentation, feature extraction, feature optimization and classification. The different combinations of CNN were discussed which are used in the implementation of automatic detection systems. The false positive reduction was taken into consideration by some of the researchers for achieving higher accuracy. However, there is further scope of improvement by inculcating patients' history, symptoms, and treatment till date with the system to get accurate results.

KEYWORDS

- computer-aided diagnosis
- convolution neural network
- medical image processing
- medical imaging
- pulmonary nodule detection
- statistical machine learning algorithms

REFERENCES

1. Hongtao, X., Dongbao, Y., Nannan, S., Zhineng, C., & Yongdong, Z., (2018). Automated pulmonary nodule detection in CT images using deep convolutional neural networks. *Pattern Recognition, 85,* 109–119.
2. Amal, E. M., Gholamreza, A., & Hooman, K., (2019). Automatic detection and segmentation of blood vessels and pulmonary nodules based on a line tracking method and generalized linear regression model. *Signal, Image and Video Processing, 13,* 457–464. https://doi.org/10.1007/s11760-018-01413-0.
3. Naman, G., Deepak, G., Ashish, K., Pedro, P. R. F., & Victor, H. C. D. A., (2018). Evolutionary algorithms for automatic lung disease detection. *Measurement, 140,* 590–608.
4. José, R. F. J., Marcelo, C. O., & De Azevedo-Marques, P. M., (2016). Integrating 3D image descriptors of margin sharpness and texture on a GPU-optimized similar pulmonary nodule retrieval engine. *J. Super Comput., 73,* 3451–3467. doi: 10.1007/s11227-016-1818-4.
5. Mohamed, S. P., Burhanuddin, M. A., & Mohamad, I. D., (2019). Lung cancer detection from CT image using improved profuse clustering and deep learning instantaneously trained neural networks. *Measurement, 145,* 702–712.
6. Lakshmanaprabu, S. K., Sachi, N. M., Shankar, K., Arunkumar, N., & Gustavo, R., (2019). Optimal deep learning model for classification of lung cancer on CT images. *Future Generation Computer Systems, 92,* 374–382.
7. Cesar, A. D. P. P., Nadia, N., & Luiza De, M. M., (2019). Detection and classification of pulmonary nodules using deep learning and swarm intelligence. *Multimedia Tools and Applications.* https://doi.org/10.1007/s11042-019-7473-z.
8. Marysia, W., & Taco, S. C., (2019). Pulmonary nodule detection in CT scans with equivariant CNNs. *Medical Image Analysis, 55,* 15–26.
9. Ratishchandra, H., Yambem, J. C., & Khumanthem, M. S., (2019). Pulmonary nodule detection on computed tomography using neuro-evolutionary scheme. *Signal, Image and Video Processing, 13,* 53–60. https://doi.org/10.1007/s11760-018-1327-4.

10. Yutong, X., Jianpeng, Z., & Yong, X., (2019). Semi-supervised adversarial model for benign-malignant lung nodule classification on chest CT. *Medical Image Analysis, 57,* 237–248.

11. Gupta, A., Das, S., Khurana, T., & Suri, K., (2018). Prediction of lung cancer from low-resolution nodules in CT-scan images by using deep features. In: *2018 International Conference on Advances in Computing, Communications, and Informatics,* 531–537.

12. Abraham, J. (2011). Reduced lung cancer mortality with low-dose computed tomographic screening. *Community Oncology 8*(10), 441–442. https://doi.org/10.1016/s1548-5315(12)70136-5.

EMBEDDING PARALLELISM IN IMAGE PROCESSING TECHNIQUES AND ITS APPLICATIONS

SUCHISMITA DAS, G. K. NAYAK, and SANJAY SAXENA

International Institute of Information Technology, Bhubaneswar, Odisha–751003, India

ABSTRACT

Image processing become demanding and attracting research domain due to its versatile application in a different field such as military field, medical imaging, authentication of the digitization process through signature recognition, face recognition, the agricultural field, etc. Due to improved technology and massive development in cost-effective image acquisition equipment, the size and number of images increase day by day. In some domains, such as medical, military applications, accuracy, as well as real-time image analysis, is the most important criterion for evaluating the performance of the system. As each application is having its own requirement, every system demands real-time, accurate, less expensive, and more extensive computation. Most of the application of computer vision, image processing, pattern recognition deploys the feature extraction to use the machine learning and deep learning methodology. All these techniques and architecture requires huge time to extract the features of the images which could not be acceptable in some of the real-world application. To reduce the time and make the computation faster and efficient, the concept of parallel processing is embedded in image analysis. This study aims to provide an introduction as well as the need of parallelism in image processing. The mostly used parallel processing using CUDA and GPU is briefly discussed with their shortcomings.

Finally, how parallel computation is used in machine learning and deep learning architecture for medical image analysis is discussed in detail to show the impact of parallel processing. Also, one case study based on brain tumor segmentation is elaborated in this chapter. The main purpose of this chapter is to summarize existing parallel image processing techniques and tools via various research and analysis and their limitation.

11.1   INTRODUCTION

Normally desktop workstation does not have abundant computational resources to support large scale image processing. Application of parallel computing techniques to image processing is well suited, as, a parallel library of common image processing tasks and capabilities of the end users to extend this library will meet certain needs. These issues led to the development of PIPT. To achieve a high degree of portability PIPT uses MPI to effect parallelism. Again, MATLAB is a commonly used high level programming language which is having a rich library and simple paradigm but it uses an interpreter in turns demeans its logic. Hence NVIDIA's CUDA is a golden revolutionary standard which allows programers to use the power of computing engine used in CUDA enabled Graphics Processing Units (GPUs).

Signal is a function which indicates how a variable change with respect to other variables. Concept of digital signal processing (DSP) is originated in the 1960–1970's [1]. But the mass cost and expensiveness of computers led to its application in limited areas. The revolution in PCs of 1970–1980's led to its entry in public domain. Images are normally two-dimensional matrix consisting of pixels; hence it is needed to be processes to reveal the hidden and unique characteristics, which led to the birth of Image processing. The images captured by different tools in different domain during these days required to be processed efficiently and time effectively with respect to their acquisition, variable size, and complexity. This brings the concept of parallel processing along with image analysis for the researchers and developers. Parallel computing using CUDA is a commonly used tools in various application in the field of image analysis such as image re-sizing, histogram equalization, thresholding, edge detection, filtering, etc. Image Processing plays a major role in optics computer science, satellite imaging, Agricultural field, medical image analysis, etc. In recent years, parallel computing is playing a major tool in high-speed computing. For implementation of this technology in real field, the several research has been done using GPU, CUDA, JAVA, Hadoop, and MATLAB. Significant speed up can be obtained by using this parallel computing based on CUDA and executed on GPU using image processing algorithms. Matrix computation and operation is generally time consuming and required massive computation. As image is a matrix with pixel values at each coordinate, the processing of images especially using machine and deep learning (DL) which makes its computation on CPU time consuming and inefficient. GPU is having tremendous accelerators which increase the speed of computation even if for regular operation on dense matrix. In Figure 11.1 it shows the basic architectural difference in CPU and GPU [2]. It makes image analysis time efficient by computing blocks of pixels in parallel for different application on data in a single program. As GPUs are having multiple number of processing cores, they can handle massive computation such as image analysis more efficiently for real world application.

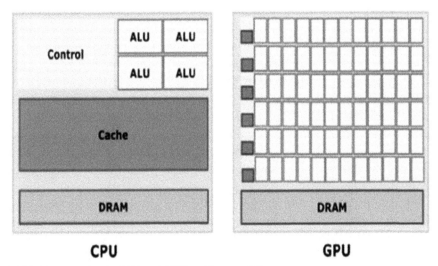

FIGURE 11.1   Basic CPU and GPU architecture [2].
Source: Reprinted from Ref. [2].

Real time image processing requires a massive extent of processing supremacy, expertise in computing to perform operations. Parallel processing of images is done to handle this issue through several tools and techniques like CUDA, GPU, PCT of MATLAB, OpenCV. In current years, parallel processing has become a significant tool to perform high speed computing. Several tools have been devised for a suitable technique of parallel computing. This study summaries existing parallel image processing techniques and tools and their benefits and limitations.

11.2 ANALYSIS OF PARALLEL IMAGE PROCESSING TECHNIQUES ON CUDA AND GPU

CUDA is a parallel computing environment developed by Nvidia for parallel high-performance computing. It is the compute mechanism in GPU and can be accessed by programer through paradigm programming languages. CUDA is solely proprietary to Nvidia video cards. Nvidia offers APIs in their CUDA SDK to give a stage of hardware extraction that hides the GPU hardware from developers; so that the developers no longer have to realize the complexities at the back of GPU. One benefit of the hardware abstraction is that it allows Nvidia to change the GPU architecture. In the following Figure 11.2, the flow of computing in CUDA is shown [3, 30]. CUDA is portable with any programming language such as C, C++ and Fortran which makes it simple, demanding, and efficient for the developers. Data is processed by GPU following the instruction given by CPU. The original data is initially loaded in main memory which is copied to GPU memory during computation. GPU processes the data in parallel in each core and the final output again copied back to the CPU main memory. Then it is available to the respective application. The original computation divided into several small parallel task and it is computed through GPU core in parallel.

Image processing and computer vision is highly demanding technique in medical image analysis for early diagnosis of different diseases. Image processing is a form of signals processing where an input is an image and the output may be an image or whatever that experiences some meaningful processing. A number of researches have been done to embed parallelism in image processing techniques [1, 4] which is summarized below:

- When medical image segmentation techniques are used such as thresholding, region growing, active contour, etc., it was pointed out that mainly image processing issues as pixels are using similar instructions and data from a miniature neighborhood around the pixel so that the thread count too high but certain segmentation methods do not process each pixel.

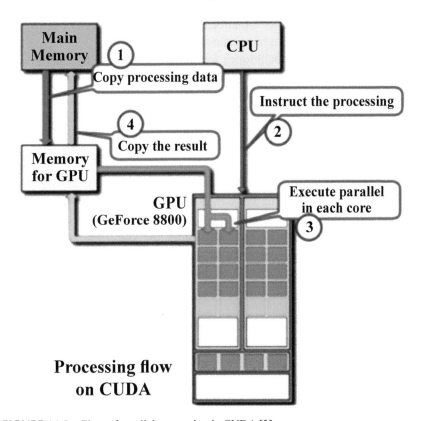

FIGURE 11.2   Flow of parallel computing in CUDA [3].

Source: Reprinted from Ref. [3]. https://creativecommons.org/licenses/by/4.0/

- When several image enhancement algorithms such as brightening filter, darkening filter, negative filter and RGB to gray scale filter are implemented then recursive ray tracing (RRT) technique is used. RRT also considers light coming through the surrounding via reflection, refraction, and shadows. It was concluded that the issues that involves high inter thread communication increase in value of number of threads per block gives faster result.
- When region growing algorithms of image processing is implemented, the pixels are considered as seeds segments and fine-grained parallel thread are assigned to individual pixels.

- Median filtering is implemented by the authors on GPU, then they have found that it is possible to gain in response time with an access level GPU, allowing real time, image, and audio filtering.
- Retinal fundus image enhancement is implemented and two custom design masks are used to enhance the image.
- In CUDA based techniques for image enhancement, Wallis transformation is used. Here, Wallis filter is used and compared with sequential implementations on CPU.
- Medical image segmentation of hepatic vascular is implemented using fuzzy connectedness method an improved algorithm for (CUDA-Kfoc) is proposed by adding a correction step on the edge points.
- A presentation of parallel implementation based on CUDA of bias field algorithms PB CFCM which is an enhanced version of FCM that accurate the in-homogeneity intensity and the image is partitioned concurrently, here firstly the centroids are initialize, then assigned and data transported from CPU to GPU before the loop iteration. Two main kernels are used, one to calculate the membership function and second one to calculate the estimated bias field.
- The performance of canny edge detection is compared. Hadoop map reduce and CUDA based satellite image processing is introduced. Significant speed up is obtained as the size of image is increased for canny edge detection.
- Here, image filtering is implemented using CUDA, where median filtering is suitable for implementation using CUDA in GPU.

High performance computing (HPC) or parallel computing has become an integral part in today's mainstream computing system. Hence GPU is pivotal for handling time consuming image processing techniques and algorithms. GPU is a processor on a graphics card to compute highly parallel calculation. Its main implementation is to transform, interpret, and quicken graphics. Instead of a CPU it contains millions of transistors specializing in arithmetic of floating point. It has revolutionized 3-D graphics revolution and empowers us to run HD graphics. While the CPU is a serial processor, the GPU is a stream processor. It makes use of control logic to accomplish the execution of many threads while maintaining resources of a sequential execution. In Figure 11.3 (taken from a lecture note on GPU architecture and CUDA programming), it shows how the CUDA performs 1D convolution operation and producing output in different core in parallel which in turns reduces the time effectively.

FIGURE 11.3   1D convolution operation in CUDA (output[i] = (input[i] + input[i+1] + input[i+2])/3.f).

Source: https://www.cs.cmu.edu/afs/cs/academic/class/15418-s18/www/

Due to CUDA and parallel programming concept, GPU accelerates and performs better. The following points are considered to represent the benefit of parallel processing through CUDA:

- CUDA is termed as Computer United Device Architecture which is a free software platform provided by Nvidia. It enables users to control GPU's by writing program as on C++. But CUDA cores tend to be more specialized for stream processing as opposed to be generalized like a CPU. Using CUDA allows the programer to take advantage of the massive parallel computing power of a Nvidia Graphics card in order to perform general purpose computation.
- In multi-core CPU and GPU, CUDA lets the programer take advantage of the hundreds of ALUs inside a graphics processor, which is much more powerful than the hundreds of ALU's available in any CPU.
- CUDA is well-suited for highly parallel algorithms.
- CUDA is best suited for number crunching.
- CUDA is well suited for large data sets.

GPUs has emerged as powerful accelerators for many regular algorithms that operate on dense vectors and matrices and it is designed to process blocks of pixels at high speed and massive parallel operations on application data like data visualizations and graphic vendering. Moreover, GPUs are designed for parallel computing based on arithmetic operations. In 2006 Nvidia [5] introduced a hardware and software platform designed for general purpose computing on GPUs, with new instruction set architecture. A CUDA capable GPU is referred to as a device and the CPU as a host. CUDA provides thread which is a fine unit of parallelism. It also provides different memory spaces which are accessible for threads during execution.

MATLAB is a very efficient and east to perform developing environment. It uses an interpreter which is slower in executing loops compared to the compilers of other languages, i.e., C, C++. As image processing tasks requires excessive use of loops, we accelerate MATLAB processing by using Nvidia's CUDA parallel processing architecture. MATLAB executable (MEX) is an important usage that allow to compile the codes written in other languages. Thus, to perform parallel computing with MATLAB and use GPU, various options are available for using GPU in MATLAB. There are some limitations of CUDA along with a number of benefits such as:

- Easy to learn CUDA API but hard to program efficient applications which use GPU performance;
- CUDA belongs to the class of SIMT;
- CUDA API is a set of extensions based on C language. This parallel architecture limitations are caused basically by HW architecture.

11.3   NEED OF PARALLELISM IN IMAGE PROCESSING

Increasing in number of images and their complexity nature requires massive computation for new technology to meet the real time deadlines with acceptable accuracy. Some image processing algorithm takes a couple of days even if a week to train the model and gives the respective output to meet certain criteria on a single processor or in sequential processing architecture. Also, in every field of computer vision, image processing and pattern recognition, complex mathematical algorithms with optimization techniques are used. To make the processing faster and to meet real time application deadline, parallel processing is required in the field of data mining, image analysis. The whole task can be distributed among several processors or the use of GPU can increase the computational time efficiently.

A typical desktop workstation does not have sufficient computational resources to support large scale image processing because of high resolution of the high resolution of medical images and imagery data. Thus, parallel computation was developed to alleviate the problems arising out of the nature of many image processing algorithms. A parallel library of common image processing tasks, as well as the capability of the end users to extend to this library was evolved to meet certain needs [1, 6]. But it had certain limitation and cannot be accommodated to a variety of users. Thus, these issues led to a development of parallel age processing kit (PIPT). The details of parallelism are hidden from users of PIPT through a registration or call back mechanism, which provides a transport mechanism for moving parameterized data between nodes. To achieve a high degree of portability, the PIPT uses the MPI standard to effect parallelism. The main implementation of MPI is in public domain so that the PIPT can make use of this standard while remaining free.

Certain computation tasks in many low level and mid-level image processing applications readily suggests a natural parallel programming approach. It will minimize the amount of data which needs to be distributed to the individual nodes and thus minimize the overall communication cost. For large scale of image processing tasks, the input image data is required to compute the certain portion of the output, i.e., localized. An output image is computed simply by independently processing single pixel of the input image. A no. of pixels is generally used to compute an output pixel. So, each individual output pixel can be computed dependently and in parallel. A high degree of natural parallelism can be used easily by parallel algorithms. Many images processing can achieve near linear setup with the addition of processing nodes.

Another challenging task in developing a parallel image processing library is making it portable among the various types of available parallel hardware. To make portable parallel coding, it is important to incorporate a model of parallelism that is used by a large number of potential targets architectures. Hence message passing the most used paradigm for the implementation of parallel programs on distributed memory architecture. Several message passing libraries [6] are available (such as P4, parallel virtual machine (PVM), PICL, and zip-code).

PIPT was designed so that it could be easily extensible by end users. New functionality can be added by registering three routines with the PIPT network. Registering functions with the PIPT creates internal accounting tables that enable the parallel framework to marshal arguments, distribute parameters and gather return values during a parallel run.

11.3.1 TASK AND DATA PARALLELISM

Data parallelism and Task parallelism are both contrast to each other. Data parallelism is the process of distributing the data between multiple processors which computes a single task on the data in parallel. Whereas task parallelism is the process of distributing concurrent tasks performed by processes or threads across different nodes. So, computing the different algorithms on same data (images) is the one way of achieving parallelism were computing the same task on different set of images is another way of it. Hence, he images analysis can be parallelized in both the way to achieve maximum speed and better performance. Task parallelism can be achieved through pipelining and the data parallelism is achieved through work sharing. There are a number of image analysis operations in which the data and task parallelism can be performed. CPUs are suitable task parallelism for some heavy processing by only a few threads whereas GPUs is more suited for applications where large amount of data parallel operations are performed. Applications can achieve optimal performance by combining data parallelism on GPU with task parallelism on CPU.

Here the point operator and neighbor operators are shown (Figure 11.4) based on data parallelism [7] where the host is distributing the data to each of the node and the operations are carried out in parallel. Finally, all the outputs are combined and synchronized to produce a single matrix or image.

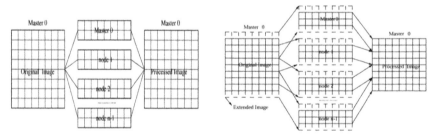

FIGURE 11.4   Data parallelism in point operation (left) and in neighborhood operation [7].

Source: Reprinted with permission from Ref. [7]. © 2002 Elsevier.

In the similar way image contrast, image histogram equalization, filtering operation, convolution operation can be done using data parallelism and hence time can be reduced. In some large image, the whole single image can be divided into smaller area [7] and each sub part of the image is fed to different nodes to processed in parallel. Diagram Figure 11.5 schematically demonstrates the global operator on a single image.

At the same time the task parallelism is required to compare the performance of different algorithms or extracting the features using different methodologies on same set of images. In Figure 11.6, the data distribution among the different task is shown. Task A and Task B both are performing different task on a single set of images. This process can also be used in a single algorithm for independent tasks. As image size and number increases in this digital world, the parallel processing is the only way to meet the criteria demanded by different applications.

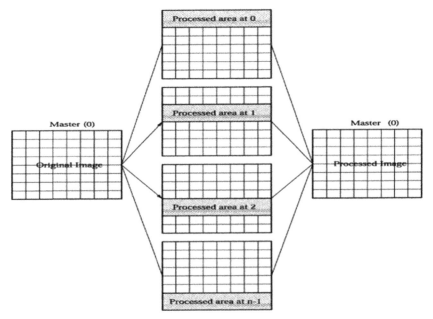

FIGURE 11.5 Data parallelism in global operation.

Source: Reprinted with permission from Ref. [7]. © 2002 Elsevier.

Author Pieter P. Jonker has done a number of experiments by taking different image size with different number of processor and plotted the graph based on the speed up. Again, he also performed the task and data parallelism on same application and different image size and noted down the speed. Form the graph it is clearly shown that the speed will be increased for different image size and remains constant after a threshold number of processes. But while observing the data and task parallelism graph, it is shown that it increases gradually and achieve more speed up than only parallel computation (Figure 11.7).

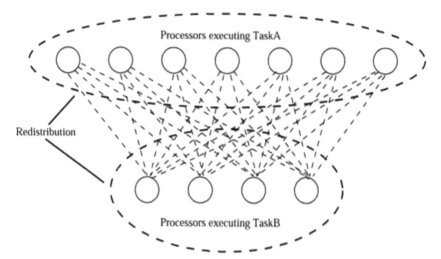

FIGURE 11.6   Task parallelism using dividing task into Task A and Task B.
Source: Reprinted with permission from Ref. [7]. © 2002 Elsevier.

11.4   EMBEDDING PARALLEL COMPUTATION IN DEEP LEARNING (DL) AND MACHINE LEARNING (ML)

Image processing and computer vision are the active research area from last few decades, but during the last few years, there is a huge and remarkable change and advances in technology in these fields due to development of machine learning (ML) and DL along with artificial intelligence (AI). Deep neural networks (DNNs) are becoming popular and attracting the attention of the researcher due to its high computational efficiency and extracting the hidden features of the images automatically. But this DNNs required large amount of data to do classification, segmentation, or recognition problems which in turn induces high computation time may be one or couple of days if trained on CPU. The huge success of the DL and ML also having limitation of training the large amount of data and having high computation time in extracting features. To address this issue and reduce the training time, the data as well as task parallelism can be implemented in the field of DL.

Processing using graphical processing unit (GPU) can reduce the training time in different image analysis application by a factor x, where x is the number of cores used in GPU. Still the computation will take order of several hours to train the DNN model (such as VGGNet, ResNet, FCNN, etc.). Hence to increase the speed and reduce the computational

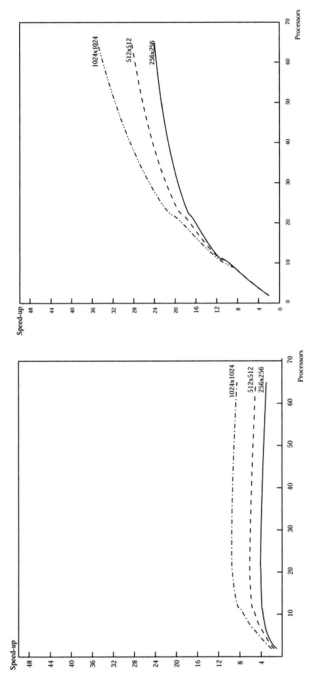

FIGURE 11.7 Speed-up for the data parallel approach (left), speed-up for the data and task parallel approach (right) [7].

time, there is a huge need of parallel and distributed algorithms which can distribute the data as well as task among several machines and cores. There are few ways in which the parallelism in image processing using DL and ML can be embedded as follows:

- Training process and data distribution: As the data set is very large now a days and also DL performs well for huge training data, it is very difficult to store the model and data as a whole in a single processor RAM. To achieve efficient and faster computation, data, and model architecture can be distributed among multiple machines. If GPU based machine having multiple cores is used then the performance will definitely increase in a great difference. The Data parallelism can be achieved by dividing the whole data into a mini batch of data and distribute it to the different machines to make the computation speeder. In very rare case the complex models are not fit into a single machine, need more memory. Hence the model can be splitted layer wise or it can be splitted forward phase and backward weight updation layer and can be fitted to different machines. If memory is a constraint for large data set or model then both can be distributed and parallel processing can be applied to make the memory efficient model.
- In ML, feature extraction and feature reduction are one of the most important steps of the algorithm. For a large data set and models, it is very time consuming and complex to extract the features manually through CPU and then applying feature selection and reduction algorithms. Evolving of Parallel processing algorithms helps a lot in reducing computational time. Here the feature extraction and reduction algorithm can be modified such that the model can perform the same task on a bunch of distributed set of images across different machines. Hence the time will be saved and the model can be trained and can be used in real time application.

Author Vishakh Hegde [8] has implemented Convolutional Neural Network in parallel. In CNN, the convolution operations are performed by sliding the filter throughout the whole image with a fixed stride size. Author in his chapter, performed each layer of convolution operation in parallel by calling the im2colgpu in parallel and computing the matrix multiplication in parallel. He has experimented by considering different batch size in each convolution layer and different matrix size in fully connected layer with both CPU and GPU and computed the average time of computation over 10 number of experimented results. The details of the performance are discussed in the chapter with graphical representation. In this chapter the parallel and sequential computation of convolution operation and matrix-matrix multiplication was performed and found that it takes huge amount of computational time, even if one week of time, on a single CPU. Doubtless, parallel processing and GPU increase the speed but at the same time, communication overhead and complexity of the algorithm and model must be investigated.

Author Tal Ben-Nun [9] has done a very intensive and useful analysis of parallel processing in DL [23] and the concurrency that can be implemented in distributed environment. In his report, he has reviewed near about 147 papers out of 240 in which detail hardware specification along with result analysis is provided in a great extent. The chapter which clearly shows the trend of using GPUs at the starting of year 2013 and even multi node trend started from year 2015 to support large computational algorithms and models. It becomes growing gradually and now a days, distributed-memory architectures with GPUs have become the ultimate option for ML and DL in each of the application. In this chapter it is very nicely shown that the parallelism in DL can be achieved through partitioning of forward training phases and backward weight updation phases. The higher computational speed can be achieved by data parallelism (performed by partitioning input samples), model parallelism (done by dividing network structure), and layer pipelining (splitting and training layer-wise).

Currently medical image application using ML, DL, transfer learning is one of the most attractive, demanding, and challenging research field due to the high rise in health problems and difficulty in diagnosis them in real time. This application demand early and on time diagnosis of the diseases based on image analysis. Another most challenging task in medical image analysis, is, images are complex in nature, variable in size, acquisition process, high resolution and vary from subject to subject. Parallel processing is embedded in medical image analysis to get the diagnosis result on time with more efficiently and accurately. There are different image modalities like MRI, CT scan, Ultra Sound images, X-ray, etc., which requires different types of per-processing and visualization methodology. Hence parallel processing can be modeled in different medical imaging application such as image reconstruction, image denoising, motion estimation, image registration, morphological operations, segmentation, and localization of organs and modeling. One of the significant ways of analyzing the complex structure inside the brain in microscopic level is provided by reconstructing the brain fiber followed by diffused weighed imaging. At the same time due to complex structure of brain and different analysis model, the process is computationally expensive. Hence many researchers have proposed parallel algorithms based on GPU to speed up the computation by a significant factor as compared to multiprocessing/ multi-threaded CPU. From last few years, different Authors in their research [14–17] have explored power of parallel computation and GPU in different medical application. Image denoising, image reconstruction on CT scan [18–21], Diffusing different models [10] in biomedical imaging on multicore processors, deformable registration for mapping brain data-sets, vertebra detection and segmentation in X-ray images are few examples of such work progresses on medical field on GPU to enhance the speed of computation. In their researches, the speed increased by a factor 10x, 20x to 40x. Not only in medical field, there is a massive computation and huge application in the remote sensing images, satellite imaging, plant diseases identification and treatment, etc. This indicates there is a huge demand and also requirement of embedding parallel computation in image analysis along with powerful distributed machine and DL algorithms.

11.5   CASE STUDY

In this section, two case studies are elaborated how the parallel processing is embedded with image processing to achieve less computation time as compare to sequential computation. This shows how the trend grows towards the parallel processing on high demand of real-world application.

In 2017, AdelArfa [11] has experimented the parallel processing and its efficiency in image filtering techniques. Image filtering using low pass filter is implemented in parallel mode using PVM and its performance, efficiency, and the speed factor is analyzed through multiprocessor machines. In the work, the data parallelism is considered to implement filtering process and to enhance the computation time. The entire image is divided into sub-images and assigned to different remotely localized machines. Each of the remote processor computes the filtering technique and send back it to the master computer where the results are combined to produce the original single filtered image. To analyze the efficiency, the different sizes of image along with different filters using a set of remotely localized distributed machines are taken and the performance is analyzed.

The process of image filtering is computed in three phases: in first phase the low pass filtering is used on an image in a sequential computer and measure its efficiency. This phase is required to compare the efficiency of the parallel processing in comparison to sequential computation. In second phase the computation is done with four slave machines and one master machine which distributed the data among the slaves [32] and manage the processing based on PVM. The architecture of the parallel processing used in this chapter is shown in Figure 11.8. Finally in last step, the analysis of the parallel computation is performed by taking different image size and low pass filter mask.

The detailed design part of the work is illustrated in Figure 11.9. It mainly goes through the 5 steps. Fist load all the images in the master machine and then divide each image into sub images. In next step the sub images are sent to the four slaves after serializing the images in master computers. Assigning the task and communication to the slaves and all kind of synchronizing is done by using PVM. Each slave computer is computing the low pass filtering process on the given sub region of images and sent back to the master processor. Once the master process receives all the filtered sub images, reassembled all to generate a single image.

The parallel computation time decreases significantly as compare to the sequential processing. It also analyzed with different image sizes. But the process is having communication overheads and the accuracy of the process depends on the proper distribution of data and synchronous communication between the multicore machines.

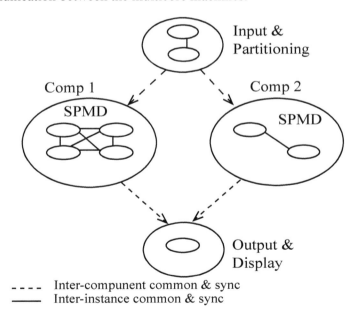

FIGURE 11.8   Parallel computation model using PVM [11].

Source: Reprinted from Ref. [11]. Open access.

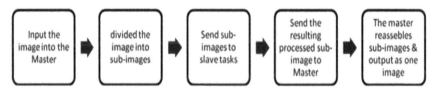

FIGURE 11.9   Flow of the work done to achieve the parallel computation [11].

Source: Reprinted from Ref. [11]. Open access.

The Parallel image segmentation program is designed using HIPI [12] which is made compatible to Hadoop 2.6.1 version. HIPI has an important feature, i.e., its integration with Open CV, which is a prominent opensource library comprising of various computer vision algorithms. HIPI has the capability to be deployed on the cluster of nodes. They have taken the input data in HIB format, i.e., HIPI image Bundles for image segmentation program which is one of the best ways of representation of large images on HDFS. HIB uses the Hadoop-Map Reduce features to its maximum which is designed to support the proper processing of large flat files. HIB class provides basic functions for reading, writing, and concatenate HIB files.

In the chapter, initial process which HIPI uses to filter out the images is called Culling. The process of culling is based on various user defined conditions, i.e., spatial dimensions or resolution associated to image meta data. The Culler class extends Map-Reduce framework and does it through HIB run time mode prior to its delivery to the mapper in a complete decoded format. After thee culling process completed, the images are assigned to the individual mappers, so that the map task is implemented and achieved maximum data locality, which sends the map-reduce code to each data node in the cluster.

The technology used in the segmentation process is scalable and can be used in a cluster of machines. But the experiments are performed on a single quad-core machine with 3.40 GHZ clock frequency and 8 GB RAM running on Ubuntu 14.04-Linux 64 bits. For task execution time of small size of image data sets (less than 257 MB), Open CV sequential mode performs better than Hadoop distributed mode which can be seen clearly from Figure 11.10. This happens because the master thread worker takes a certain fixed amount of time slot to process each chunk of image data sequentially without any thread context switching whereas in Hadoop distributed mode, the factor of split size comes delay in task execution. The input split size set for the proposed experiment is 126 MB, which is the most compatible split size with Hadoop 2.6.1 version to obtain the maximum output.

It is found that the image segmentation done using sequential programming has a relatively stable CPU core usages which averages around 135 over the entire execution. But the theoretical CPU core usages is 14%. The 1% difference is due to the I/O disk usage operation. The image segmentation is implemented sequentially is totally cache bound. If the application wants to write to the memory location, then there might arise a cache issue for other cores [33–36].

11.6 CONCLUSION

This chapter summarizes the need of parallelism in the field of image processing and how the application of GPU increases the performance of the DL and ML based model. It is best suited for most of the image processing algorithms [31] such as image filtering, image registration, classification shows parallel behavior. Data parallelism and task parallelism both can be implemented efficiently in image processing techniques with the help of GPU and CUDA programming. With the help of parallel and distributed algorithm, the massive computation time of images decreasing significantly and meet the real time application criteria. Parallelism in image processing can be achieved [13] in both software and hardware level through the different hardware architecture and parallel algorithm. But at the same time, it also incurs the communication overhead and requires synchronous communication to get the accurate output. It also induces the hardware complexity in the model. Some factors like synchronization, sharing memory, branching, and dividing the data/task efficiently also limit the speed of the parallel model architecture. Different GPU optimization also required to overcome this limitation somehow.

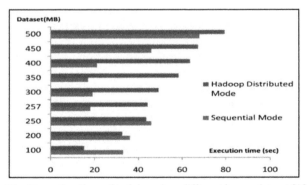

FIGURE 11.10   Execution time of task based on different image dataset [12].

Source: Reprinted from Ref. [12]. Open access.

As the research in DNN [29] is growing rapidly in the field of image processing due to its efficient and accuracy in comparison to most of the ML algorithm, the time as well as memory becomes the constraint in the computation of the DL models. To do efficient computation and to use optimized memory in handling the DNN sparsity, the NVIDIA Tensor Cores [26], the tensor processing unit (TPU) [27], other ASICs and FPGAs [24, 25], and even neuromorphic computing [28] are used now a days in many research papers. Due to the current situation and huge demand of real time application, Parallel processing dominates the sequential processing in many applications, at same time it is also having hardware and resource limitation which requires attention for the researchers.

KEYWORDS

- branching
- deep neural networks
- graphics processing units
- MATLAB executable
- parallel virtual machine
- recursive ray tracing

REFERENCES

1. Squyres, J., & Lumsdaine, A., & Stevenson, R., (1999). A toolkit for parallel image processing. *Proc. SPIE, 3452.* 10.1117/12.323475.
2. Horrigue, L., & Ghodhbane, R., & Saidani, T., & Atri, M., (2018). *GPU Acceleration of Image Processing Algorithm Based on MATLAB CUDA, 18*(6), 91–99.
3. Hangun, B., & Eyecioğlu, Ö., (2017). Performance comparison between open CV built in CPU and GPU functions on image processing operations. *International Journal of Engineering Science and Application, 1*(2), 34–41.
4. Sanjay, S., Shiru, S., & Neeraj, S., (2017). Study of parallel image processing with the implementation of VHGW algorithm using CUDA on NVIDIA'S GPU framework. *Proceedings of the World Congress on Engineering, I.*
5. Zlatanov, N., (2016). *CUDA and GPU Acceleration of Image Processing.* 10.13140/RG.2.1.4801.0004.
6. Saxena, S., & Sharma, S., & Sharma, N., (2016). Parallel image processing techniques, benefits and limitations. *Research Journal of Applied Sciences, Engineering and Technology, 12*, 223–238. 10.19026/rjaset.12.2324.

7. Nicolescu, C., Jonker, P. A data and task parallel image processing environment, Parallel Computing, Volume 28, Issues 7–8, 2002, 945-965, https://doi.org/10.1016/S0167-8191(02)00105-9.

8. Hegde, V., & Sheema, U., (2016). *Parallel and Distributed Deep Learning.* pp. 1–8.

9. Ben-Nun, T., & Torsten, H., (2019). Demystifying parallel and distributed deep learning: An in-depth concurrency analysis. *ACM Computing Surveys.* https://doi.org/10.1145/3320060.

10. Khor, H. L., Liew, S., & Zain, J. M., (2015). A review on parallel medical image processing on GPU. In: *2015 4th International Conference on Software Engineering and Computer Systems* (pp. 45–48). Kuantan. doi: 10.1109/ICSECS.2015.7333121.

11. AdelArfa, R. J., (2017). Parallel processing implementation using PVM for image processing filtering. *American Journal of Engineering Research, 6*(12), 209–219.

12. Akhtar, M. N., Saleh, J., & Grelck, C., (2018). Parallel processing of image segmentation data using Hadoop. *International Journal of Integrated Engineering, 10.* 10.30880/ijie.2018.10.01.012.

13. Hemnani, M., (2016). Parallel processing techniques for high performance image processing applications. In: *2016 IEEE Students' Conference on Electrical, Electronics and Computer Science* (pp. 1–4). Bhopal. doi: 10.1109/SCEECS.2016.7509316.

14. So, H., Chen, J., Yiu, B., & Yu, A. (2011). *Medical Ultrasound Imaging: To GPU or Not to GPU? IEEE Micro, 31*(5), 54–65. https://doi.org/10.1109/mm.2011.65.

15. Evert, V. A., Neda, S., Andrei, J., & Anna, V., (2011). CUDA-accelerated geodesic ray-tracing for fiber tracking. *International Journal of Biomedical Imaging, 2011*, 12. Article ID 698908. doi: 10.1155/2011/698908.

16. Fabian, L., Sidi, A. M., Mohammed, B., Saïd, M., & Pierre, M., (2011). Heterogeneous computing for vertebra detection and segmentation in x-ray images. *International Journal of Biomedical Imaging, 2011*, 12. Article ID 640208. doi: 10.1155/2011/640208.

17. Meilian, X., & Parimala, T., (2011). Mapping iterative medical imaging algorithm on cell accelerator. *International Journal of Biomedical Imaging, 2011*, 11. Article ID 843924. doi: 10.1155/2011/843924.

18. Martin, S., (2011). GPU-accelerated finite element method for modeling light transport in diffuse optical tomography. *International Journal of Biomedical Imaging, 2011*, 11. Article ID 403892. doi: 10.1155/2011/403892.

19. Daehyun, K., Joshua, T., Mikhail, S., Clifton, H., Pradeep, D., & Armando, M., (2011). High performance 3D compressive sensing MRI reconstruction using many-core architectures. *International Journal of Biomedical Imaging, 2011*, 11. Article ID 473128. doi: 10.1155/2011/473128.

20. Anders, E., Mats, A., & Hans, K., (2011). True 4D image denoising on the GPU. *International Journal of Biomedical Imaging, 2011*, 16. Article ID 952819. doi: 10.1155/2011/952819.

21. Daniel, J. T., Can, C., Anthony, K., et al., (2011). Patient specific dosimetry phantoms using multichannel LDDMM of the whole body. *International Journal of Biomedical Imaging, 2011*, 9. Article ID 481064. doi: 10.1155/2011/481064.

22. D'Amore, L., Daniela, C., Livia, M., & Almerico, M., (2011). Numerical solution of diffusion models in biomedical imaging on multicore processors. *International Journal of Biomedical Imaging, 2011*, 16. Article ID 680765. doi: 10.1155/2011/680765.

23. LeCun, Y., Bengio, Y., & Hinton, G., (2015). Deep learning. *Nature, 521*(7553), 436–444.

24. Chen, T., Du, Z., Sun, N., Wang, J., Wu, C., Chen, Y., & Temam, O., (2014). DianNao: A small-footprint high-throughput accelerator for ubiquitous machine learning. In: *Proc. 19ᵗʰ Int'l Conf. on ASPLOS* (pp. 269–284).

25. Nurvitadhi, E., et al., (2017). Can FPGAs beat GPUs in accelerating next-generation deep neural networks?. In: *Proc. ACM/SIGDA International Symposium on Field-Programmable Gate Arrays (FPGA '17)*, 5–14.

26. NVIDIA, (2017). *Programming Tensor Cores in CUDA, 9*. https://developer.nvidia.com/blog/programming-tensor-cores-cuda-9/ (accessed on 3 November 2021).

27. Jouppi, N. P., et al., (2017). In-datacenter performance analysis of a tensor processing unit. In: *Proc. 44th Annual International Symposium on Computer Architecture (ISCA '17)* (pp. 1–12).

28. Akopyan, F., et al., (2015). Truenorth: Design and tool flow of a 65 mW 1 million neuron programmable neurosynaptic chip. *IEEE Transactions on Computer Aided Design of Integrated Circuits and Systems, 34*(10), 1537–1557.

29. Chan, W., Jaitly, N., Le, Q., & Vinyals, O., (2016). Listen, attend and spell: A neural network for large vocabulary conversational speech recognition. In: *IEEE Int'l Conf. on Acoustics, Speech and Signal Processing* (pp. 4960–4964).

30. CUDA Zone, (2020). Retrieved from https://developer.nvidia.com/cuda-zone (accessed on 30 September 2021).

31. Wu, A. Y., (2001). Parallel image processing. In: Davis, L. S., (ed.), *Foundations of Image Understanding* (Vol. 628). The Springer international series in engineering and computer science. Springer, Boston, MA.

32. Netlib, (2000). *The PVM System* [Online]. Available from: http://www.netlib.org/pvm3/book/node17.html (accessed on 30 September 2021).

33. Saxena, S., Sharma, N., & Sharma, S., (2015). A significant approach for implementing parallel image processing using MATLAB with java threads and its implementation in segmentation using Otsu's method in multi core environment. *International Journal of Applied Engineering and Research, 10*(17), 37651–37657.

34. Saxena, S., Sharma, N., & Sharma, S., (2015). Parallel computing in genetic algorithm (GA) with the parallel solution of n queen's problem based on GA in multicore architecture. *International Journal of Applied Engineering and Research, 10*(17), 37707–37716.

35. Saxena, S., Sharma, N., & Sharma, S., (2015). Parallel computing in genetic algorithm (GA) with the parallel solution of n queen's problem based on GA in multicore architecture. *International Journal of Applied Engineering and Research, 10*(17), 37707–37716.

36. Saxena, S., Sharma, N., & Sharma, S., (2016). Parallel image processing techniques, benefits and limitations. *Research Journal of Applied Science, Engineering & Technology, 12*(2), 223–238.

CHAPTER 12

HIGH-PERFORMANCE COMPUTING AND ITS REQUIREMENTS IN DEEP LEARNING

BISWAJIT JENA, GOPAL KRISHNA NAYAK, and SANJAY SAXENA

International Institute of Information Technology, Bhubaneswar, India, E-mail: c118002@iiit-bh.ac.in

ABSTRACT

Modern-day computing involves the computation of complex, multi-dimensional and volumetric data—the generation of these data, basically from the web world, networking, and high-end applications. Natural images, medical images, hyperspectral images, and video datasets are always in complex form. To handle these data and run the application that generates these kinds of data and especially solve the complex problems related to these datasets, high-performance computing is the ultimate choice. The recent success and trending of deep neural networks for solving computer vision applications is a big leap in the computation direction. So, to solve complex computer vision problems with the deep neural network, high-performance computing is again the good choice, as deep neural networks are with deep architecture and many more trainable parameters. In this chapter, we cover many more aspects of high-performance computing and its requirements in deep learning in a competitive approach to help the potential researchers furthers.

12.1 INTRODUCTION

With the increase in data in the modern day of computation and the world of the internet, networking, there is a huge requirement for high-power computational machines. Then there is a need for the evolution of high-performance computing machines to handle these volumetric data. The data may be text, numbers, tables, files, audio, image, and video, or a combination of these. The forms in which data are available not always simple as we need. Data are available in complex forms such as biomedical images, satellite images, hyperspectral images, surveillance images, and video. The data generated from engineering applications, medical industry, production firms are always huge, volumetric, complex, and multi-dimensional in nature. To study, analyze, process, and get a better statistical outcome from these data, we need a high-performance computing machine for high-performance computing (HPC) [1, 2].

We can define high-performance computing (HPC) simply because it uses multiple computers to solve bigger tasks faster than the traditional computing approach. It uses a parallel programming structure rather than sequential programming. So, HPC is a cluster of systems working together seamlessly as one unit to achieve the performance goals such as processing data and performing complex calculations at high speeds.

Image data and video data in the form of a natural image, satellite image, medical image, and surveillance video always hold special weightage in this modern computing and internet world. Again these data produced by different applications are large in size. A deep learning framework as a form of high-performance computing has been very efficient in computing these image and video data. Deep learning networks [3–5], in these recent times, have achieved huge success and are hence efficient for image processing tasks. Due to its capacity to solve complex problems, deep learning technologies workload is growing faster than ever before. The current work in the area of deep learning technologies is a big leap in the field of artificial intelligence. The deep neural network is a very revolutionary step in computer vision, image classification, and speech recognition. The pre-trained models of deep learning technologies are giving good help in this direction.

High-performance computing and its requirement in deep learning [6, 7] are the most discussed topics in recent times. The intersection of HPC and Artificial Intelligence (AI) [8] based computation (machine learning and deep learning) provide more cutting edge computation power to virtually unlimited data. Deep learning or data analysis requires high computational power for specific applications that lead to high-performance computing innovation. Because AI systems are designed to process so much data, it needs to run on optimized hardware that has the capacity to perform trillions of calculations per second or more. This is where HPC and AI converge since HPC utilizes dense computer clusters in sync with one another to run the most advanced AI. The converse Venn diagram of these technologies is depicted in Figure 12.1. Throughout the long term, developments in CPU and GPU equipment combined with achievements in disseminated programming have permitted HPC to at last make profound learning a practical undertaking. Before the development of HPC, it was challenging to train a simple neural network. However, with the invention of the deep neural network, it is possible within a few hours and even in a few minutes. This has likewise raised HPC, close by AI, as an essential resource for present-day organizations.

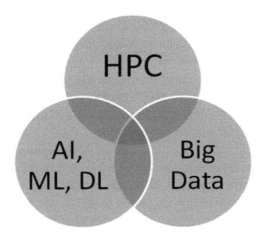

FIGURE 12.1   Convergence of HPC, AI, and Big Data.

The developments and more suggest that high-performance computing is the next frontier for enterprise AI, and many large tech companies are grabbing the opportunity to facilitate these developments.

The remaining parts of the chapter are organized as follows. Section 12.2 describes high-performance computing and its importance. Section 12.3 illustrates various popular parallel programming models. Section 12.4 introduces the various aspects of GPU and CUDA. Artificial Intelligence, Machine Learning, and Deep Learning features and technologies are then described in Section 12.5. Section 12.6 lists out the various pre-trained model and transfer learning. Section 12.7 explains Deep Learning – A huge time complex computing technique. The need for parallel computing in deep learning technologies is described in Section 12.8. Section 12.9 presents the CUDA deep models and Conclusion in Section 12.10.

12.2   HIGH-PERFORMANCE COMPUTING

High-Performance Computing (or HPC) [1, 2] is the terminology in the computation field where many supercomputers are connected to accomplish many tasks, particularly to solve complex and advanced problems with phenomenal algorithms at greater speed. Reasonably, most HPCs fall somewhere close to workstations and supercomputers. Be that as it may, while supercomputers are likewise HPCs, not all HPCs are supercomputers. To achieve the tasks under HPC, many technologies are brought together, such as electronics, computer architecture, programs, system software, and algorithms working under a single canopy to produce fast, cost-effective, efficient results of the kind that are unachievable by regular computers. HPC systems use computing resources concurrently to deliver sustained performance. HPCs are a key element in handling today's demanding computer workloads, from Big Data to predictive analytics to machine learning and artificial intelligence.

We use supercomputers in the case of high-performance computing compared to general-purpose computers and exploit the features, properties, and characteristics of supercomputers. The performance of general-purpose computers is measured in MIPS, which stands for million instructions per second. With supercomputers, performance is instead measured in FLOPS, which stands for floating-point operations per second. A supercomputer is defined as one that can far outperform general-purpose computers in terms of speed, reliability, efficiency, and problem-solving capacity. There are supercomputers that have the capability to perform up to around a hundred quadrillion FLOPS, with the majority of the fastest using Linux as their operating system. A high-performance computing system does not necessarily contain any components that you would not find in a general-purpose computer. The difference is mainly in quantity, as HPCs are composed of computing clusters configured to work together. Whereas a general-purpose computer typically contains a single processor, a supercomputer contains several processors, each comprised of anywhere between two and four cores. Each individual computer within an HPC cluster is referred to as a node, so a supercomputer with 64 nodes may have up to 256 cores, all working in tandem. When high numbers of individual nodes work efficiently together, they can often solve problems that would be too complex for a single computer to solve by itself.

12.2.1   HPC ARCHITECTURE AND IMPLEMENTATION

The biggest challenge in the HPC system lies in its implementation. The specialty of the HPC is its unique architecture. It works in parallel fashion, i.e., instead of a single monolithic computer to perform all computation, HPC is a set of computers doing parallel computation. It forms a cluster of machines to achieve high computation in which the machines are linked to each other in a parallel fashion. The machines inside the cluster are known as "nodes." The numbers of nodes inside the cluster basically depend upon the size of the cluster. The multiple nodes inside a cluster exploit the parallelism of the cluster. The main computing unit of the node is the processor inside the node. These processors may be multi-core CPUs (central processing units) or GPGPUs (general-purpose graphics processing units). The CPUs are considered the brain of computers, which has ALU (Arithmetic Logic Unit) to perform all arithmetic and logical operation. GPGPUs are a general-purpose version of GPUs, which process graphics and images. The machine will be more parallelism if the number of cores is more inside the processor. Multi-core processors, like multiple clusters, enable parallel processing [9, 10]. HPC architecture connects a multi-core processor with an HPC node, and HPC clusters are shown in Figure 12.2.

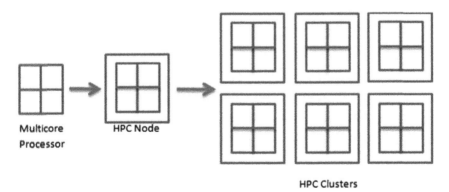

FIGURE 12.2   HPC architecture connects a multi-core processor with an HPC Node and an HPC Cluster [9].

HPCs, as different sorts of PCs, need more than hardwires to run. They need a working framework that empowers clients to use applications. HPCs utilize either a Linux or Windows working framework. Linux is a group of free and open-source programming working frameworks dependent on the Linux piece. Windows is Microsoft's business working framework. Linux will be more famous for elite figuring than Windows since Linux is a clone of the UNIX working framework, and supercomputers utilize the UNIX working framework. HPC systems require a low-latency network with high bandwidth so that the individual nodes and clusters can communicate and connect with one another. High-performance computing is used across multiple disciplines, including climate modeling, geographical data analysis, biosciences, electronic design automation, modeling for the oil and gas industry, and entertainment and the media. Small to medium businesses may also use some form of high-performance computing, using a cluster as small as four nodes or 16 cores. Despite the smaller size, this type of HPC can still help solve problems faster and more efficiently than a single general-purpose computer.

The principal advantages of HPC are cost, adaptable organization model, speed, adaptation to non-critical failure, and absolute expense of proprietorship, despite the fact that the genuine advantages acknowledged can change from framework to framework. Numerous ventures use HPC for critical thinking, including life sciences, assembling, and oil and gas, and more use cases are showing up over the long run. Some examples include Fraud prevention, Drug discovery, genome mapping, Celestial body discovery, Live streaming, Video special effects many more [9].

12.3 PARALLEL PROGRAMMING MODELS

A parallel programming model [11, 12] exists as a set of programming abstractions for fitting parallel activity applications to the underlying parallel model hardware. The two important aspects that judge the parallel programming model are the generality of the architecture that fits into numerous ranges of applications and the performance of the parallel programming model. The commonly used parallel programming models are: (a) Shared memory model (b) Message passing model, (c) Thread model (d) Data parallel model, and (e) Hybrid model.

12.3.1 SHARED MEMORY MODEL

The shared memory model [13, 14] under the parallel programming model states that the memory (common address space) has been shared among the processors that are accessed by them asynchronously. The data and the program available in this shared memory are accessed by the processes under various processors and start executing their individual parts. After completing the execution processes, all the processors are rejoined back to the main program to provide the output. The pictorial representation of this concept is depicted in Figure 12.3.

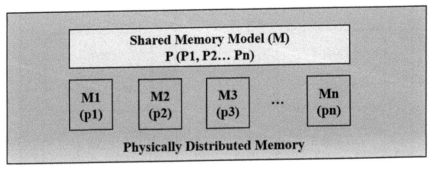

FIGURE 12.3   The structural representation of shared memory concept.
Source: Modified from Ref. [12].

Deadlock and race conditions are the main concern for the kind of program under the shared memory concept [15, 16]. Deadlock is a situation where the two more processes cannot progress further as they are in turn waiting for some other processes to finish, which in turn are again waiting for some other processes to finish and so on. Furthermore, the race condition is a situation where two or more processes try to access and modify the same data almost simultaneously. The final results might not be the desired results in this case. However, multi-core processors support shared memory concepts with languages and libraries like OpenMP.

12.3.2 MESSAGE PASSING MODEL

The message passing [17, 18] model works with passing messages from one process to another. The process with its own set of data may reside under the same processor or with a different processor. The co-operation between every process while data transfer between the processes is achieved by send ( ) and receive ( ) function. There must be a receive operation for every send operation. There are mainly two different message passing modes: synchronous and asynchronous. Synchronous message passing includes passing messages between two processes where the receiver must be ready to receive the message. In an asynchronous mode, the receiver is not ready to receive the message. The illustration of this message passing system is shown in Figure 12.4.

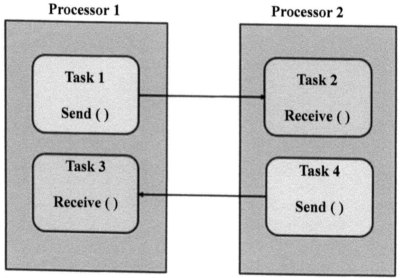

FIGURE 12.4   The system of message passing of parallel programming model.
Source: Modified from Ref. [12].

12.3.3   THREAD MODEL

The thread programming model [19, 20] is a type of shared-memory programming. A single "heavyweight" process can have multiple "lightweight," concurrent execution paths in the threads model of parallel programming. A thread can be defined as a short sequence of instructions within a process. A different number of threads can be executed on the same processor or on different processes. If the threads are executed on the same processor, the processor switches between the threads randomly. If the threads are executed on different processors, they are executed simultaneously, which is shown in Figure 12.5.

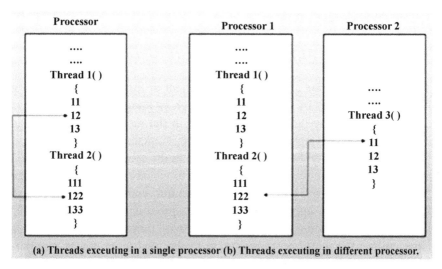

(a) Threads exceuting in a single processor (b) Threads executing in different processor.

FIGURE 12.5   Thread model representation of parallel programming.

Source: Modified from Ref. [12].

12.3.4   *DATA-PARALLEL MODEL*

Data parallelism [21, 22] is one of the simplest forms of parallelism. Here data set is organized into a standard structure such as an array. Many programs apply the same operation to different parts of the typical structure. Suppose the task is to add two arrays of 100 elements store the result in another array. If there are four processors, then each processor can make 25 additions.

12.3.5   *HYBRID MODEL*

A hybrid model [23–25] combines more than one previously described programming model. In these types of models, any two or more parallel programming models are combined. They are developed based on specific application-oriented. A typical example of a hybrid model is the message passing model (MPI) with the threads model (OpenMP). Another similar and increasingly popular example of a hybrid model is using MPI with CPU-GPU (Graphics Processing Unit) programming. Other hybrid models are standard: MPI with Pthreads and MPI with non-GPU accelerators.

12.4 GPUS AND CUDA

12.4.1 *GRAPHICAL PROCESSING UNIT (GPU)*

The GPU has become an essential component of modern-day computing with the availability of volumetric data such as image, video, audio, and other forms of these data. GPUs are used in the modern computing age in every personal computer, laptop, desktop, workstations, mobile phone, game console, and embedding system as a multi-core and multi-threaded multi-processor. The fields of study related to visual processing like image processing, computer graphics, and computer vision prominently use GPU to process its applications. The high processing capacity of GPU is being credited by heavily parallel processing units [26, 27]. The GPUs have ignited the world of computation to the next level, hence popularising Artificial Intelligence (AI) with becoming a crucial part of modern supercomputing.

12.4.2 *CPU vs. GPU vs. TPU*

The central processing units (CPUs) have been largely used because of their popularity to solve multiple computing applications hassle-freely. It was initially designed for computing purposes in all kinds of applications involving arithmetic and logic operation. It has been used mostly on laptops, desktops, embedded systems, and mobile devices to run software and application of various ranges. The CPU also supports various kinds of operating systems of 32 bit or 64bit, such as Linux, Windows, Mac OS, and all kinds of real-time operating systems (RTOS). Even if the CPU (Central Processing Unit) can process visual data computing, it still has some limitations. While GPUs (Graphical Processing Unit) have thousands of cores present in them, CPUs have limited. Hence CPUs are good for serial processing, whereas GPUs are preferred for parallel processing. In the case of GPU, various digital signal processing operations are performed to make the addition, subtraction, multiplication, division, and high-end applications like gaming, high-definition video editing, streaming, etc. It also needs some extra hardware to perform these applications. TPU (Tensor Processing Unit) [28–30] is a customized central processor worked by Google to do huge loads of information preparation at low accuracy. It is utilized by numerous individuals of the applications which google offers like google search, google map or google photographs. The best thing about TPU is its Artificial Intelligence features, such as machine learning and deep learning. Likewise, it isn't ready for selling purposes now, and different TPUs are associated with the structure of a supercomputer to crunch a colossal measure of information that google creates from its server.

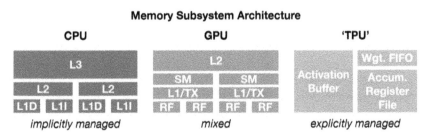

FIGURE 12.6   Memory subsystem architecture of CPU, GPU, and TPU [28].

Source: Reprinted from Ref. [28]. Open access.

In summary, we recommend CPUs for their versatility and for their large memory capacity. GPUs are a great alternative to CPUs when you want to speed up a variety of data science workflows, and TPUs are best when you specifically want to train a machine learning model as fast as you possibly can. Figure 12.6 shows the memory subsystem architecture difference between these processors, and Figure 12.7 illustrates the TPU architecture.

12.4.3   THE GPU ARCHITECTURE

The NVIDIA company coined the first GPU architecture in late 1990. The company's GeForce range of graphics cards was the first to be popularised and ensured related technologies could evolve. Later, AMD and Intel were the two main players in the graphics card arena. The GPU consists of multiple processing clusters, each having multiple streaming microprocessors. These SMs accommodate multiple threads, multiple processor cores, an L-1 cache layer, and an L-2 shared cache. Each processor core inside the SM executes instructions for parallel threads. To address different market segments, GPUs are often scaled by the number of processor cores and memories. This enables them to use the same architecture and software. In the case of NVIDIA, this is done by varying the number of streaming multi-processors and DRAM [32, 33]. Figure 12.8 shows the schematic architecture of NVIDIA's GPU.

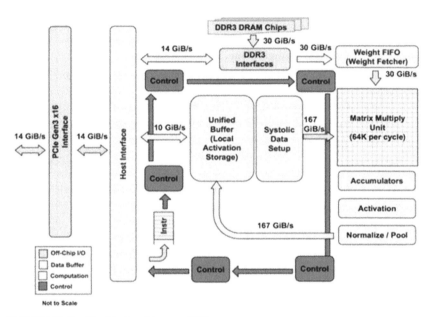

FIGURE 12.7   The TPU architecture [29].

Source: Reprinted from Ref. [29].

12.4.4   THE CUDA PROCESSING

The CUDA programming model [32–34] allows developers to exploit that parallelism by writing natural, straightforward C code that runs in thousands or millions of parallel invocations or threads. To utilize the GPU, CUDA provides a symbolic C-like programming language and compiling. Any call to a global function is said to issue a kernel running on the GPU and specify the dimension of the grid and the blocks used to execute the program. Now, Figure 12.9 will depict the example of some basic features of parallel programming with CUDA. It contains sequential and parallel implementations of the SAXPY routine defined by the basic linear algebra subroutines (BLAS) library.

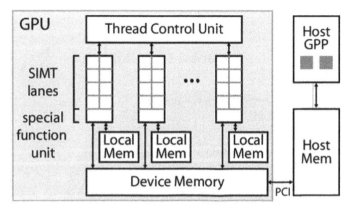

FIGURE 12.8 GPU Schematic Architecture [31].

Source: Reprinted from Ref. [31]. Open access.

CPU Code	CUDA 6 Code with Unified Memory
```void sortfile(FILE *fp, int N) {```	```void sortfile(FILE *fp, int N) {```

```
void sortfile(FILE *fp, int N) {
    char *data;
    data = (char *)malloc(N);

    fread(data, 1, N, fp);

    qsort(data, N, 1, compare);

    use_data(data);

    free(data);
}
```

```
void sortfile(FILE *fp, int N) {
    char *data;
    cudaMallocManaged(&data, N);

    fread(data, 1, N, fp);

    qsort<<<...>>>(data,N,1,compare);
    cudaDeviceSynchronize();

    use_data(data);

    cudaFree(data);
}
```

FIGURE 12.9 CPU code vs. CUDA code with unified memory.

12.5 ARTIFICIAL INTELLIGENCE, MACHINE LEARNING, AND DEEP LEARNING

High-performance computing is closely related to Artificial intelligence (AI) and hence to Machine learning (ML) and Deep Learning (DL), as we discussed in the introduction section that. Artificial Intelligence is the study of computer science in which artificial intelligence is induced to machines, which will be treated as the natural intelligence of a human. AI is a broad domain under which ML and DL are a subset which is depicted in Figure 12.10.

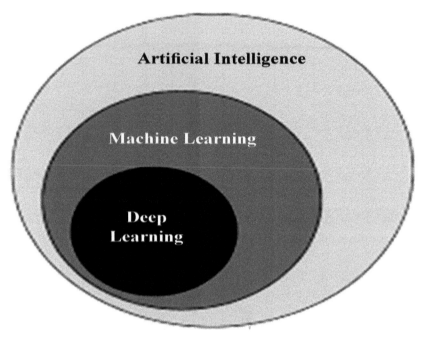

FIGURE 12.10   AI vs. ML vs. DL.

12.5.1   ARTIFICIAL INTELLIGENCE

Artificial Intelligence [8] is generally creating an application running on an algorithm that can take decisions or processes depending on the input provided. In a much simpler sense, if you see any application such as recognition of handwriting, spam classification, text summarization, etc. Although there might be a common perception that artificial intelligence is the same as machine learning, it is very important to understand that they are not the same as Machine Learning is the subset of Artificial Intelligence and Deep Learning is the subset of Machine Learning. Artificial intelligence involves other major concepts such as Reinforcement Learning, natural language processing, autonomous vehicles, etc. Especially natural language processing and autonomous vehicles are those kinds of fields where most of the research is application-oriented, and there is high freedom of choosing various approaches to solve the problem. Statistical machine learning, traditional machine learning, and deep learning, which involves the major utilization of calculus derivatives for optimization of parameters, will need much more information regarding the above-mentioned terms in the upcoming subsections.

As we have discussed what artificial intelligence is, we will now discuss the subsets of artificial intelligence that are the most profound topics that sparked the interest of various researchers across different fields. The first subset we will discuss is Machine Learning. Machine Learning is automated learning by computer where a dataset is given as input which consists of various features, and the system predicts the output. The most common subdivisions of it are Supervised Learning and Unsupervised Learning, and there is also semi-supervised learning which is discussed later.

12.5.2 MACHINE LEARNING

Machine learning [35] is mainly divided into Supervised Learning, which means that learning involves the categorization of input data. Supervised Applications can be found in day-to-day life, such as spam filters, digit recognition written by hand, famous dog or cat classification, and so on. If we observe carefully, the dataset of each mentioned application involves features; for example, in a movie review system, the star rating given by the user can be the feature if the user gives a more rating, the user review can be taken as positive, and if a low rating is given it is taken as a negative review, here we can clearly see the input formalism which is the rating and the output formalism which classification as positive review or negative review. In such cases, the machine learning model requires the dataset to include rating and also classification of each rating as positive or negative while training the model. Hence, it requires a human to label the dataset into categories which is really tiring. However, the advancement in Deep learning and the recent introduction of generative adversarial networks proved to be very beneficial for generating samples, especially for better efficiency without requiring huge amounts of the dataset. Discussion of generative adversarial networks is out of the scope of this book but strongly recommended for people interested in computer vision, natural language processing, etc. Another type of subdivision of machine learning is unsupervised machine learning, where there is no need to label the dataset or the prediction does not involve categorizing each individual sample into a predefined category. A general example of unsupervised learning is regression analysis which involves starting with a predefined hypothesis which is generally a polynomial in parameters and data features, and we need to optimize the parameters using a given dataset; in a more general sense in the dataset, we divide each sample into a sequence of input feature vector and output value which we need to predict. Our goal is to learn the general function of mapping the area of land to the price of the land. For this, let us suppose that the true mapping function 'f(x) = y,' where x represents the area of the land and y denotes the price of the corresponding area of the land. Here the function is a simple linear function using linear regression that is the hypothesis at the start is considered to be linear as an example, and the possible hypothesis can be:

$$h(x) = \Theta * x + c \qquad (1)$$

Our goal is now to optimize the theta as it is an unknown parameter as well as constant parameter 'c.' We can use various optimization algorithms to do so, but the most famous one is the gradient descent algorithm, where we take the derivative of the mean squared error between the hypothesis's predicted value and true value in the dataset and optimize the theta accordingly. Discussion of various regression algorithms and classification algorithms is out of the scope, and it is strongly recommended for all individuals to go through the various regression algorithms and classification algorithms. It is very much important to discuss another important application of unsupervised learning which generally involves an arbitrary dataset, and we need to find the group of samples which are similar to each other, for five instances, if you are given customer data of shopping mall, then based on their expenditure you can group them into various categories as most visiting, least visiting and so on which are not evident from looking at the data itself especially when the dataset is too huge, In such scenarios clustering algorithms such as K-means, hierarchical clustering, etc. are helpful. Clustering algorithms are of profound importance as they possess the power to group the samples of the data into clusters and are sometimes beneficial in creating supervised learning datasets to deal with supervised learning tasks.

12.5.3 DEEP LEARNING

Deep Learning [3–5] is an alternate approach to solving various machine learning tasks using neural network architecture. First, it is beneficial to see what a neural network is as it is the fundamental building block of the concept. A neural network is a graph inspired by the functioning of neurons of the human body, where the information is received at the starting layer of neurons, and it is "forward-propagated" till the output layer. Simple neural network architecture can be visualized, as shown in Figure 12.1, where we can see an input layer going through a series of layers until the final output layer is reached. To better understand this, we can assume the input vector to be an "n x 1" dimensional, implying that the dataset we are working on has 'n' features. We take a weight matrix containing weights that need to be optimized according to the relation between the previous and current layers. In other words, what we are doing here is calculating the current layer values based on the previous layer and weight matrix; this procedure is called forward propagation. A simple forward propagation formula for the neural network is:

$$Z^L = (W^L * Z^{L-1}) + b \qquad (2)$$

In Eq. (2), "W^L" denotes the weight matrix of dimension "m x n" where "m" is the number of neurons in layer "L" and "n" is the number of neurons in layer "L-1." Z^i denotes the column vector of dimension equaling the number neurons in that layer and values calculated through the forward propagation equation. "bL" denotes the bias values that need to be added as offset while estimation of layer "L" values.

As illustrated in Figure 12.11, there is an input layer of n dimension, a hidden layer of n dimension including the bias unit, and an output layer of dimension 2. Hence, two weight matrices will be required for forwarding propagation with dimensions "3 x n" and "1 x n" from left to right. In forwarding propagation, while calculating the current layer values bias unit is not calculated from the input layer as it is the constant or offset parameter in hypothesis estimation. The mentioned information only covers forward propagation but does not cover backpropagation which is the algorithm for updating weights in weight matrices. Explaining backpropagation in a generalized way can be confusing; hence, just consider Figure 12.11. As in Eq. (2), the neuron values are calculated, but, in practice, we generally activate the values with activation functions; that is, we apply a sigmoid or tanh or relu function to the layer to make a model learning much more efficient. Hence, while implementing backpropagation, we take this fact also into consideration. Now, let us assume the activation function for the hidden layer is g (z), where z is the vector of dimension '4' as calculated using Eq. (2). We generally find the error between the output unit and the true values, which is:

$$J \text{ (output layer)} = \text{output layer} - \text{true values} \qquad (3)$$

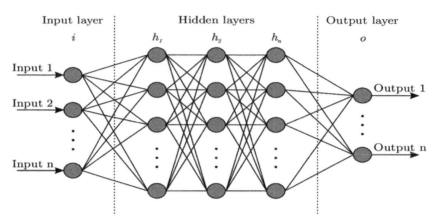

FIGURE 12.11   A simple neural network [36].

Source: Reprinted from Ref. [36]. © 2021 Elsevier. https://creativecommons.org/licenses/by/4.0/

Now, we need to translate this error so as to calculate the error of the previous layer, which is the hidden layer, which is:

$$J \text{ (hidden layer)} = ((W^2)^T * J \text{ (output layer)}) \times g(z^1) \qquad (4)$$

Note: '*' in Eq. (4) represents matrix multiplication, whereas 'x' represents element-wise multiplication.

When we have finally calculated errors of layers now we need to update the weights which is:

$$W^2 = (W^2 + (J \text{ (output layer)} * g(z^1))) \qquad (5)$$

$$W^1 = (W^1 + (J \text{ (hidden layer)} * X)) \qquad (6)$$

Note: X denotes the feature vector.

Finally, after looking at the complex backpropagation algorithm, the complete simple neural networks have been completed. This basic idea is further developed as Convolutional Neural Networks in which specified layers like convolutional layer and pooling layers and CNN is widely used in computer vision applications. One can always see how it changed even the field of NLP, and to move on further, we see recurrent neural networks which are widely used for natural language processing applications such as machine translation, speech recognition, time-series processing, etc., where advanced concepts like LSTM is used which is an alternate version for RNN. They perform much better than RNN for the above-specified applications. The next subsection will be a briefing on reinforcement learning which will be the final part discussed in this section.

12.5.4   REINFORCEMENT LEARNING

Reinforcement learning [37] is what the robots are programmed with and to capture the entire subject in a small space is literally tough; hence, a basic outline of what it is very much important to truly understand semi-supervised learning. Unlike in the above-discussed subsections, there are no categories to classify or there is anything to predict; here, we need to give a "reward" to the robot to increase or decrease the probability of a certain action it takes. The robot may visit the current state or may not visit it again, but we need to specify the rewards for each action that can be taken in a particular state. Thus, as we are not exactly saying which action to take at a specific state or giving instructions about which path to follow, and we are providing it with only rewards based on which the robot has to act, it becomes semi-supervised. Consider a dog, when you throw a ball, if it comes back picking up a stick, you give a reward, but when it picks up the ball and comes back, you give it a reward, and it will understand that whenever the ball is thrown, it should get the ball. This simple idea is a very basic example of how reinforcement learning in robots works, and there are many more algorithms in reinforcement learning, including deep learning algorithms such as Deep Q-Learning. One can always find interdisciplinary research going with reinforcement learning and other fields of machine learning.

12.6   PRE-TRAINED MODELS OF DEEP LEARNING

Pre-trained models [38] can be effective if you are already working on an application that involves nearly identical tasks or aligns with the state-of-the-art models. State-of-the-art models are those which have been already implemented by various institutes and made a mark with their accurate and efficient results. These models come in handy as most of the established models are open-source, and they help us save time and effort while building many of the crucial steps of a deep learning model, which are architecture, initialization of weights, selection of optimizers, diverse training dataset, efficient weights of weight matrices or optimal hyperparameters and so on. Generally, if the application is highly complex, then an individual can save a lot of time but achieve a state of the art results with minimal effort, and thus this concept becomes very much important for the people who want to implement deep learning concepts practically and want to obtain results which are highly precise and accurate. For example, assume a company has trained a deep learning model for image recognition and has achieved a state of the art results and also made it open-source. Now, if you are interested in tasks such as face recognition or image classification, etc. then, you can use the parameters of the image recognition model but slightly modify it to fit your necessities and can obtain nearly accurate results for your application. To discuss all of the pre-trained models in detail is out of scope, but what are the most common and important concepts for implementing pre-trained models is really important, and definitely transfer learning is one important thing that needs to be understood as it is the backbone for the implementation of pre-trained models. There will also be a briefing on established few pre-trained models, which are really interesting to know.

12.6.1 TRANSFER LEARNING

Transfer learning [39] is a most powerful concept in the area of deep learning applications as anyone can achieve state-of-the-art results if the task is in alignment with the established pre-trained models. The way it works is simple, but it takes some conditions to achieve efficient results. Coming to the definition of the word "Transfer Learning," it is simply making use of knowledge gained through training from scratch by the model in a particular application and transferring it to another model in close alignment so that there is no reason for the user to train the model from scratch and make it capable of producing exactly same or similar results with accuracy. As we have seen the meaning of the term, now it will be good if we are able to understand the working by taking an example. Consider Figure 12.12 (a) and let us assume that an individual has performed training on using the same architecture, then, let us assume that another individual wants to perform a task which is in close relation with the application of Figure 12.12(a) and hence slightly modifies the architecture as in Figure 12.12 (b), which is, by eliminating the output unit and adding two more layers including the output layer. Now, if the individual wants to perform the training, he can keep the weight matrices of Figure 12.12(a) architecture part, or better known as the pre-trained part, as constant and only change parameters with respect to the error propagated by backpropagation or choose to completely use the architecture and the

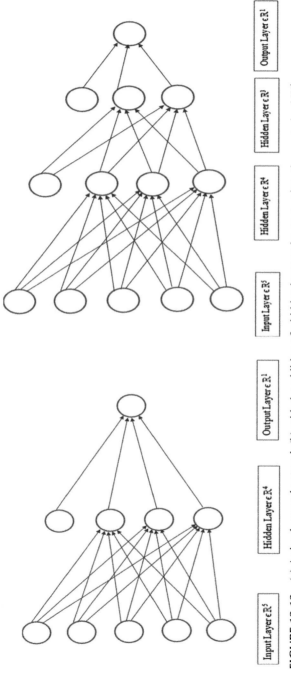

FIGURE 12.12 (a) A simple neural network (b) with the addition of a hidden layer and output unit to the pre-trained W^1.

pre-trained weights as initialization. In either of the cases, the procedure is correct, but it depends on certain conditions on when to use the first case or second case, which is discussed later. As for now, we need to understand the equation in terms of forwarding propagation in both cases. Forward propagation for the first case is as follows:

Up to pre-trained layers forward propagation equation is:

$$Z^L = (W_c^L * Z^{L-1}) + b^L \qquad (7)$$

From the pre-trained part to the appended part output layer, the equation is,

$$Z^L = (W_t^L * Z^{L-1}) + b^L \qquad (8)$$

Note; "W_c^L" is the weight matrix for calculation of Z^L and is kept constant no backpropagation happens, whereas, for "W_t^L," the weights are updated as we allow backpropagation to happen. Hence in the simple sense, for L ϵ [Pre-trained hidden layer (1), pre-trained hidden layer (1)], the weights are kept unchanged, and for remaining layers, the weight matrices are allowed to change.

Now, to discuss the second case, including the update of weights even to the pre-trained part, is a meager extension of case 1 where the backpropagation is also extended to the pre-trained weight matrices. Thus, by doing this, one can get efficient results personalized to the task or application. This marks the complete fundamental understanding of Pre-trained models in the sense of implementation and the concept of transfer learning, which marks its importance when it comes to applying these models. One should also know some important possibilities as to when to freeze the weights or free the weights, for which the answer is very simple, it depends upon how much your application is aligning with the pre-trained model and how much amount of dataset you have, that is, if there is much of an alignment with the pre-trained model you can freeze the weights but if it isn't, then, you can use the pre-trained weights as initialization and free the freezing of weights. Similarly, if you have a huge amount of dataset to train upon, then one can free the weights and leave them to backpropagation to train, and if you have a minimal dataset and the application is in major alignment with the pre-trained model, then you can freeze the weights and train. If transfer learning is interesting, then one can always go to an application known as style transfer which is more or less intuitive when one understands transfer learning.

Transfer learning indicates the jump in the model's performance when it uses transfer learning, and it also indicates a higher level of asymptotic level performance compared to a normal model while performing the same task. Figure 12.12 shows how a pre-trained model is used by using it as an initial part of a deep learning model and how an extension part is added to it for the model's training according to the given application. This concludes the information regarding transfer learning which is much about fundamentals is and it is highly recommended to pursue further knowledge on this as it is revolutionizing many fields. Figure 12.13 displays an example model of transfer learning.

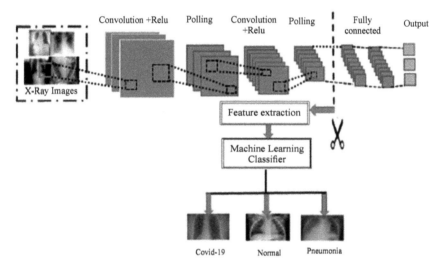

FIGURE 12.13 An example model of transfer learning.

12.6.2 SOME NOTABLE PRE-TRAINED MODELS

Transfer learning has to be used on a pre-trained model, but some important pre-trained models are what this section is all about. The pre-trained models are the successful deep learning model for image classification, object detection, speech recognition, image segmentation, and many more applications. This section is a briefing about some notable pre-trained models of deep learning.

Here is the list of some well-known *image classification* pre-trained models.

1. LeNet-5
2. AlexNet
3. VGG (VGG 16 and VGG 19)
4. ZFNet
5. GoogLeNet (Inception-v1, v2, v3 and ResNet)
6. MobileNet
7. ResNet-50
8. ResNeXt
9. Xception
10. DenseNet
11. SqueezNet
12. SENet
13. NASNet
14. PolyNet
15. PyramidNet
16. ShuffleNet
17. DPN (Dual Path Network)
18. NIN(Network in Network)
19. Maxout
20. Highway
21. SSPNet
22. PReLU-Net
23. STN (Spatial Transformer Network)
24. DeepImage
25. FractalNet
26. Pre-Activation ResNet
27. RiR (ResNet in ResNet)
28. RoR (ResNet of ResNet)
29. WRN (Wide Residual Network)
30. Stochastic Depth
31. DRN (Dialted Residual Network)
32. DPN (Dual Path Network)
33. MSDNet(Multi-scale Dense Network)
34. Trimps-Soushen
35. Residual Attention Network

Here is the list of some well-known *object detection* pre-trained models

1. OverFeat

2. R-CNN
3. Fast R-CNN
4. Faster R-CNN
5. YOLO-v1 (You Only Look Once), YOLO-v2 and YOLO-v3
6. MR-CNN and S-CNN
7. DeepID-Net
8. Craft
9. R-FCN
10. ION (Inside Outside Network)
11. MultipathNet
12. NoC
13. Hikvision
14. GDB Net, GDB-v1 and GDB-v2
15. G-RMI
16. TDM (Top down modulation)
17. SSD (Single Shot Detector) and DSSD (Deconvolutional SSD)
18. FPN (Feature Pyramid Network)
19. RetinaNet
20. DCN (Deformable Convolutional Network)

Here is the list of some well-known *image segmentation* pre-trained models.

1. FCN (Fully Convolutional Network)
2. CFS-FCN (Coarse-to-Fine Stacked Fully Convolutional Net)
3. U-Net
4. 3D U-Net
5. V-Net
6. SegNet
7. DeConvNet
8. DeepLabv1, DeepLabv2, and DeepLabv3
9. CRF-RNN
10. ParseNet
11. DilatedNet
12. DRN (Dilated Residual Network)
13. RefineNet
14. GCN (Global Convolutional Network)
15. PSPNet
16. CUMedVision1, CUMedVision2

17. Multichannel
18. U-Net+ ResNet
19. SDS (Simultaneous Detection and Segmentation)
20. Hypercolumn
21. Deepmask
22. Sharpmask
23. Multipathnet
24. MNC (Multi-task Network Cascade)
25. FCIS (Fully Convolutional Instance-aware Semantic Segmentation)

Here is the list of some well-known *Super Resolution* pre-trained models.

1. SRCNN
2. FSRCNN
3. VDSR
4. ESPCN
5. RED-Net
6. DRCN
7. DRNN
8. LapSRN
9. MS-LapSRN
10. SRDenseNet

Here is the list of some well-known *Natural Language Processing (NLP)* pre-trained models.

1. OpenAI's GPT-2 and GPT-3
2. Google's BERT
3. Google's ALBERT
4. Microsofts's CodeBERT
5. ELMo
6. Flair
7. Trnasformer-XL and XLNet
8. ULMFiT
9. Facebook's RoBERTa
10. StanfordNLP

Inception architecture is known to be one of the most creative architectures in computer vision as its applications range from image recognition to object detection and proved its impact in natural language processing applications, mainly in the design of transformer machines classification tasks, etc. Bert model is a known name in many natural language processing-related fields, which proved its importance mainly in tasks such as text processing, text cleaning, word embedding, machine translation, and other transformer-related applications. It is famous for its state of the art results and ease of use in various applications, and not to forget XL Net, an equally known natural language processing tool. Other important models are Alexnet, VGG16, Mobilenet, etc., which are computer vision-related pre-trained models but are also effective in interdisciplinary research. With this one, we can get the idea of what a pre-trained model is and how effective it can be when provided an opportunity for cross-platform knowledge. U-Net, Segnet and DeconvoNet are three important pre-trained models of image segmentation. U-Net was basically developed for biomedical image segmentation and hence good for all medical images. SegNet is an encoder-decoder-based segmentation technique that has its own advantages and is good for basically natural image segmentation. DeconvoNet follows the same architecture as SegNet however, there are fully connected layers which makes the model larger.

12.7   DEEP LEARNING: A HUGE TIME COMPLEX COMPUTING TECHNIQUES

As we discussed in the various sections, deep learning is a subfield of machine learning concerned with algorithms inspired by the structure and function of the brain called artificial neural networks. We can call deep learning a large neural network. As modern-day computation demands, it is the combination of volumetric data, a bigger model, and more computation.

As the name suggests, the deep neural network can be a huge network of many (deep) layers. We can dive deeper into the network by adding more layers into the network. Generally, the deep learning model programmer makes it deep as per the requirement of the application. To solve big complex problems with the bigger dataset, which may have many dimensions, we need a big complex network structure. Again, the deeper the network will be, the better feature it can learn from the dataset and predict better results.

Deep Learning has become the hottest growing technology in the last decade. It is now being used in various industrial sectors to make tasks simpler and grow in business. One such technology is still being explored to find out what more can be done using Deep Learning.

Healthcare industries use Deep Learning to detect diseases such as tumors, cancer cells, and diabetic retinopathy. Deep Learning has high usage in Logistics to find shorter routes to deliver the goods. This helps in saving time and fuel. It is widely used for making driverless vehicles and for automated voice generation. Deep Learning also plays a role in music generation, earthquake prediction, picture colorization, etc. The gaming industry also uses Deep Learning to build games for mobile and desktops. It is majorly used by E-Commerce and online entertainment companies to give recommendations to customers. So, from the above examples, you can see that Deep Learning is used in each and every business. But, Deep Learning comes with its own challenges. One of the major challenges is to train vast volumes of data using clusters of CPUs and GPUs. Sometimes, it takes days to train a network. Emerging techniques like transfer learning and GANs help to overcome this challenge. To wrap this, Deep Learning comes with its own pros and cons. It all depends on how you want to use it.

12.8 NEED FOR PARALLEL COMPUTING TECHNIQUES IN DEEP LEARNING

Nowadays, in this industrial world, the industry's applications are critical in nature and need heavy computation. For personal or small use cases, you can get by using non-parallel processing, but for a critical application, we have only a choice of parallel processing. Deep neural networks usually have millions or more parameters to train. It means that to get proper parameters, you need a large amount of data to go through a slow process of stochastic gradient descent type of forward-backward process to gain accurate prediction power a little bit at a time. All of these are time-consuming and computationally expensive. Parallel processing on a cluster of machines or GPU comes to help shorten the time. It's essential to large-scale deep learning.

Profound neural networks are acceptable at finding relationship structures in information in a solo style. Accordingly, it is broadly utilized in discourse examination, normal language preparing, and PC vision. In regular profound neural organizations, there are 1,000,000 boundaries that characterize the model and require a lot of information to get familiar with these boundaries. This is a computationally serious cycle that takes a ton of time. Normally, it takes the request of days to prepare a profound neural organization like a VGG network on a solitary center CPU. Once in a while, the informational index is too huge to even think about being put away on a solitary machine. Hence it is essential to concoct equal and dispersed calculations, which can run a lot quicker and which can radically decrease preparing times [41].

Even if we start from the very basic laws of parallel computing, i.e., Amdahl's law states the use of parallel processors over serial processors to increase the speed up of execution while dealing with a heavy amount of data. Figure 12.14 above shows that with an increase in processor, the speedup of execution grows exponentially. So, the deep learning network, which generally deals with volumetric data with large numbers of parameters to train the network, needs more processors in a parallel fashion. One of the beauties of a deep learning network is to deal with complex and heavy datasets. This is another reason to choose parallel computing for deep learning also.

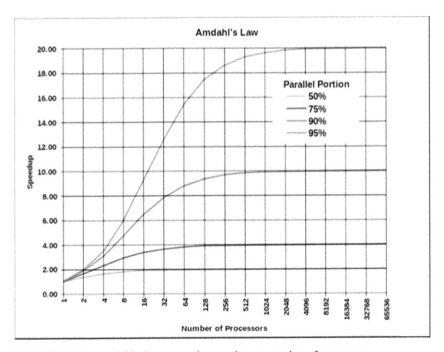

FIGURE 12.14 Amdahl's law states the speedup vs. number of processors.

Source: Image by Daniels220. https://creativecommons.org/licenses/by-sa/3.0/

12.9   A CASE STUDY: CUDA IN DEEP MODEL

CUDA (Computer Unified Device Architecture) is NVIDIA's parallel computing architecture that enables dramatic increases in computing performance by harnessing the power of the GPU.

Even if we can implement some standard and modified Deep learning architectures without CUDA on normal CPU-based machines, having the GPU processor and working on it using CUDA makes things much faster. Frameworks like Theano help in that direction. The research scientist is working on harnessing the power of GPUs to accelerate a machine learning algorithm, which has not yet been optimized for GPU commercially, and then yes, you need to learn CUDA. For instance, there are several algorithms, including but not limited to, distributed-training and parallel gradient descent, which are provided by state-of-the-art frameworks such as Tensorflow. Similar to these algorithms, you can write your own algorithm using CUDA to harness the computing power of GPU if the said idea/algorithm is not implemented yet. This approach is applicable to engineers working in industries too. The project might require you to tweak the underlying CUDA implementation. If so, it's a no-brainer that you need to understand how CUDA works and the concepts of GPU programming.

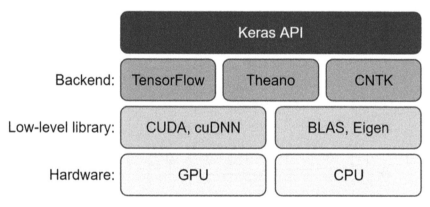

FIGURE 12.15   CUDA and Deep Learning Libraries [40].

Source: Reprinted from Ref. [40].

The relationship between the deep learning libraries and CUDA complements each other for heavy computation networks of deep learning. The

NVIDIA CUDA Deep Neural Network library (cuDNN) is a GPU-accelerated library of primitives for deep neural networks. cuDNN provides highly tuned implementations for standard routines such as forward and backward convolution, pooling, normalization, and activation layers. Figure 12.15 shows the level of relationship between CUDA and deep learning libraries. The lowermost layers are the hardware structure such as GPU and CPU. Next to it, the low-level libraries to support the hardware programming are CUDA, cuDNN, BLAS, and Eigen. Then to support neural network-based deep learning, the backend libraries are Tensorflow, Theano, and CNTK, and Keras on top of these as frontend libraries.

12.10   CONCLUSION

This chapter explores high-performance computing and its requirement in deep learning with architecture and application in depth. We also discuss all the necessary tools and terminologies related to HPC and deep learning for further enhancing the knowledge of the reader and developer. The overall study suggests that high-performance computing and its requirement in deep learning helps to save the time of computation and solve larger problems. So, high-performance computing with deep learning technologies is fast and the future of computing. We can incorporate all the current trends and technologies in the high-performance computing study make it the most inter-disciplinary study to provide state-of-the-art performance.

KEYWORDS

- **artificial intelligence**
- **deep learning**
- **graphical processing unit**
- **high-performance computing**
- **linear algebra subroutines**
- **machine learning**
- **real-time operating systems**

REFERENCES

1. Dowd, Kevin, & Charles Severance. (2010). *High-Performance Computing*.
2. Jackson, Keith R., et al. (2010). "Performance analysis of high-performance computing applications on the amazon web services cloud." *2010 IEEE Second International Conference on Cloud Computing Technology and Science*. IEEE, 2010.
3. Goodfellow, Ian, et al. (2016). *Deep Learning*. Vol. 1. Cambridge: MIT Press.
4. LeCun, Yann, Yoshua Bengio, & Geoffrey Hinton. (2015). "Deep learning." *Nature, 521*(7553), 436–444.
5. Gao, Xiaohong W., Rui Hui, & Zengmin Tian. (2017). "Classification of CT brain images based on deep learning networks." *Computer Methods and Programs in Biomedicine, 138*, 49–56.
6. Elnaggar, Ahmed, et al. (2020). "ProtTrans: Towards Cracking the Language of Life's Code Through Self-Supervised Deep Learning and High-Performance Computing." *arXiv preprint arXiv:2007.06225*.
7. Mahmoudi, Sidi Ahmed, et al. (2020). "Multimedia processing using deep learning technologies, high-performance computing cloud resources, and Big Data volumes." *Concurrency and Computation: Practice and Experience*, e5699.
8. Nilsson, Nils J. (2014). *Principles of Artificial Intelligence*. Morgan Kaufmann.
9. https://www.datamation.com/data-center/high-performance-computing.html (accessed 4 November 2021).
10. Li, Min, & Yisheng Zhang. (2009). "Hpc cluster monitoring system architecture design and implement." *2009 Second International Conference on Intelligent Computation Technology and Automation. Vol. 2. IEEE*.
11. Barney, Blaise. (2010). "Introduction to parallel computing." *Lawrence Livermore National Laboratory 6*(13), 10.
12. Zaccone, Giancarlo. Python parallel programming cookbook. Packt Publishing Ltd, 2015.
13. Adve, Sarita V., & Kourosh Gharachorloo. (1996). "Shared memory consistency models: A tutorial." *Computer, 29*(12), 66–76.
14. Gharachorloo, Kourosh, et al. (1990). "Memory consistency and event ordering in scalable shared-memory multi-processors." *ACM SIGARCH Computer Architecture News 18* (2SI), 15–26.
15. Chandra, Rohit, et al. *Parallel Programming in OpenMP*. Morgan Kaufmann, 2001.
16. Dagum, Leonardo, & Ramesh Menon. (1998). "OpenMP: an industry-standard API for shared-memory programming." *IEEE Computational Science and Engineering, 5*(1), 46–55.
17. Gropp, William, et al. (1999). *Using MPI: Portable Parallel Programming with the Message-Passing Interface. Vol. 1*. MIT Press.
18. Athas, William C., & Charles L. Seitz. (1988). "Multicomputers: Message-passing concurrent computers." *Computer, 21*(8), 9–24.
19. Riener, Andreas. (2010). *Sensor-Actuator Supported Implicit Interaction in Driver Assistance Systems*. Wiesbaden: Vieweg+ Teubner.

20. Wilson, Andrew, & Nuria Oliver. (2005). "Multimodal sensing for explicit and implicit interaction." *11th International Conference on Human-Computer Interaction (HCI International 2005), Las Vegas, Nevada, USA.*

21. Foster, Ian. (1994). "Task parallelism and high-performance languages." *IEEE Concurrency 3*, 27–36.

22. Subhlok, Jaspal, et al. (1993). "Exploiting task and data parallelism on a multicomputer." *Proceedings of the Fourth ACM SIGPLAN Symposium on Principles and Practice of Parallel Programming.*

23. Peyton Jones, Simon, et al. (2008). "Harnessing the multi-cores: Nested data parallelism in Haskell." *IARCS Annual Conference on Foundations of Software Technology and Theoretical Computer Science.* Schloss Dagstuhl-Leibniz-Zentrum für Informatik.

24. Bik, Aart, J. C., & Dennis B. Gannon. (1997). "Automatically exploiting implicit parallelism in Java." *Concurrency: Practice and Experience 9*(6), 579–619.

25. Alexandrov, Alexander, et al. (2015). "Implicit parallelism through deep language embedding." *Proceedings of the 2015 ACM SIGMOD International Conference on Management of Data.*

26. Keckler, Stephen W., et al. (2011). "GPUs and the future of parallel computing." *IEEE Micro 31*(5), 7–17.

27. Kirtzic, J. Steven, Ovidiu Daescu, & Richardson, T. X. (2012). "A parallel algorithm development model for the GPU architecture." *Proc. of Int'l Conf. on Parallel and Distributed Processing Techniques and Applications.*

28. Chen, T., et al. (2018). TVM: End-to-End Optimization Stack for Deep Learning. DeepAI. https://deepai.org/publication/tvm-end-to-end-optimization-stack-for-deep-learning.

29. Hemsoth, N. (2017). First In-Depth Look At Google's Tpu Architecture. The Next Platform. https://www.nextplatform.com/2017/04/05/first-depth-look-googles-tpu-architecture/.

30. Jouppi, Norman, et al. (2018). "Motivation for and evaluation of the first tensor processing unit." *IEEE Micro 38*(3), 10–19. (accessed 4 November 2021).

31. Linderman, Michael D., et al. (2010). "High-throughput Bayesian network learning using heterogeneous multi-core computers." *Proceedings of the 24th ACM International Conference on Supercomputing.*

32. Nickolls, John, & William J. Dally. (2010). "The GPU computing era." *IEEE Micro 30*(2), 56–69.

33. Garland, Michael, et al. (2008). "Parallel computing experiences with CUDA." *IEEE Micro 28*(4), 13–27.

34. Kirk, David. (2007). "NVIDIA CUDA software and GPU parallel computing architecture." *ISMM. Vol. 7.*

35. Alpaydin, Ethem. (2020). *Introduction to Machine Learning.* MIT Press.

36. Bre, Facundo, Juan M. Gimenez, & Víctor D. Fachinotti. (2018). "Prediction of wind pressure coefficients on building surfaces using artificial neural networks." *Energy and Buildings 158*, 1429–1441.

37. Sutton, Richard S., & Andrew G. Barto. (2018). *Reinforcement Learning: An Introduction.* MIT Press.

38. Marcelino, Pedro. (2018). "Transfer learning from pre-trained models." *Towards Data Science.*

39. Zhou, Joey Tianyi, et al. (2014). "Hybrid heterogeneous transfer learning through deep learning." *Proceedings of the National Conference on Artificial Intelligence.*
40. Pandey, P. My Journey into Deep Learning using Keras: An introduction to building and training a neural network from scratch in Keras. Towards Data Science. https://towardsdatascience.com/my-journey-into-deeplearning-using-keras-part-1.
41. Hegde, Vishakh, & Sheema Usmani. (2016). "Parallel and distributed deep learning." *Tech. Report, Stanford University.*

INDEX

Milton Keynes UK
Ingram Content Group UK Ltd.
UKHW021527161024
449569UK00059B/15